T0239995

Advanced Data Analytics Using Python

With Architectural Patterns, Text and Image Classification, and Optimization Techniques

Second Edition

Sayan Mukhopadhyay
Pratip Samanta

Apress®

Advanced Data Analytics Using Python: With Architectural Patterns, Text and Image Classification, and Optimization Techniques

Sayan Mukhopadhyay
Kolkata, West Bengal, India

Pratip Samanta
Kolkota, West Bengal, India

ISBN-13 (pbk): 978-1-4842-8004-1
https://doi.org/10.1007/978-1-4842-8005-8

ISBN-13 (electronic): 978-1-4842-8005-8

Managing Director, Apress Media LLC: Welmoed Spahr
Acquisitions Editor: Celestin Suresh John
Development Editor: James Markham
Coordinating Editor: Mark Powers
Copyeditor: Kim Wimpsett

Cover designed by eStudioCalamar

Cover image by David Clode on Unsplash (www.unsplash.com)

Distributed to the book trade worldwide by Apress Media, LLC, 1 New York Plaza, New York, NY 10004, U.S.A. Phone 1-800-SPRINGER, fax (201) 348-4505, e-mail orders-ny@springer-sbm.com, or visit www.springeronline.com. Apress Media, LLC is a California LLC and the sole member (owner) is Springer Science + Business Media Finance Inc (SSBM Finance Inc). SSBM Finance Inc is a **Delaware** corporation.

For information on translations, please e-mail booktranslations@springernature.com; for reprint, paperback, or audio rights, please e-mail bookpermissions@springernature.com.

Apress titles may be purchased in bulk for academic, corporate, or promotional use. eBook versions and licenses are also available for most titles. For more information, reference our Print and eBook Bulk Sales web page at www.apress.com/bulk-sales.

Any source code or other supplementary material referenced by the author in this book is available to readers on GitHub (github.com/apress). For more detailed information, please visit www.apress.com/source-code.

Printed on acid-free paper

The reason for the success of this book is that it has original research, so I dedicate it to the person from whom I learned how to do research: Dr. Debnath Pal, IISc.

—Sayan Mukhopadhyay

Table of Contents

About the Authors

Sayan Mukhopadhyay has more than 13 years of industry experience and has been associated with companies such as Credit Suisse, PayPal, CA Technologies, CSC, and Mphasis. He has a deep understanding of applications for data analysis in domains such as investment banking, online payments, online advertising, IT infrastructure, and retail. His area of expertise is in applying high-performance computing in distributed and data-driven environments such as real-time analysis, high-frequency trading, and so on.

He earned his engineering degree in electronics and instrumentation from Jadavpur University and his master's degree in research in computational and data science from IISc in Bangalore.

Pratip Samanta is a principal AI engineer/researcher with more than 11 years of experience. He has worked for several software companies and research institutions. He has published conference papers and has been granted patents in AI and natural language processing. He is also passionate about gardening and teaching.

About the Technical Reviewer

 **Joos Korstanje** is a data scientist with more than five years of industry experience in developing machine learning tools, of which a large part is forecasting models. He currently works at Disneyland Paris where he develops machine learning for a variety of tools.

Acknowledgments

Thanks to Labonic Chakraborty (Ripa) and Soumili Chakraborty.

Introduction

We are living in the data science/artificial intelligence era. To thrive in this environment, where data drives decision-making in everything from business to government to sports and entertainment, you need the skills to manage and analyze huge amounts of data. Together we can use this data to make the world better for everyone. In fact, humans have yet to find everything we can do using this data. So, let us explore!

Our objective for this book is to empower you to become a leader in this data-transformed era. With this book you will learn the skills to develop AI applications and make a difference in the world.

This book is intended for advanced user, because we have incorporated some advanced analytics topics. Important machine learning models and deep learning models are explained with coding exercises and real-world examples.

All the source code used in this book is available for download at `https://github.com/apress/advanced-data-analytics-python-2e`.

Happy reading!

CHAPTER 1

A Birds Eye View to AI System

In this book, we assume that you are familiar with Python programming. In this introductory chapter, we explain why a data scientist should choose Python as a programming language. Then we highlight some situations where Python may not be the ideal choice. Finally, we describe some best practices for application development and give some coding examples that a data scientist may need in their day-to-day job.

OOP in Python

In this section, we explain some features of object-oriented programming (OOP) in a Python context.

The most basic element of any modern application is an *object*. To a programmer or architect, the world is a collection of objects. Objects consist of two types of members: *attributes* and *methods*. Members can be private, public, or protected. *Classes* are data types of objects. Every object is an instance of a class. A class can be inherited in child classes. Two classes can be associated using *composition.*

Python has no keywords for public, private, or protected, so *encapsulation* (hiding a member from the outside world) is not implicit in Python. Like C++, it supports multilevel and multiple inheritance. Like Java, it has an `abstract` keyword. Classes and methods both can be abstract.

© Sayan Mukhopadhyay, Pratip Samanta 2023
S. Mukhopadhyay and P. Samanta, *Advanced Data Analytics Using Python*,
https://doi.org/10.1007/978-1-4842-8005-8_1

In the following code, we are describing an object-oriented question-answering system without any machine learning. The program's input is a set of dialogs in input.txt, as shown here:

```
glob is I
prok is V
pish is X
tegj is L
glob glob Silver is 34 Credits
glob prok Gold is 57800 Credits
pish pish Iron is 3910 Credits
how much is pish tegj glob glob ?
how many Credits is glob prok Silver ?
how many Credits is glob prok Gold ?
how many Credits is glob prok Iron ?
how much wood could a woodchuck chuck if a woodchuck could
chuck wood?
Program has a knowledge base in config.txt.
I,1,roman
V,5,roman
X,10,roman
L,50,roman
C,100,roman
D,500,roman
M,1000,roman
```

Based on this input and the configuration program, the answer to the question is given in input.txt in standard output, as shown here:

```
pish tegj glob glob is 42
glob prok Silver is 68 Credits
glob prok Gold is 57800 Credits
glob prok Iron is 782 Credits
I have no idea what you are talking about
```

The parsing logic is in the Observer class.

```python
import operator

#this class verify the validity of input
class Observer(object):
    #store frequecy of symbols
    length = {}
    #most frequent symbol
    symbol = ''
    #count of most frequent symbol
    count = 0
    #calling class
    compiler = None

    def __init__(self,cmpiler):
        self.compiler = cmpiler

    def initialize(self, arr):
        for i in range(len(arr)):
            self.length[arr[i]] = 0

    #increase count for each occurence of symbol
    def increment(self,symbol):
        self.length[symbol] = self.length[symbol] + 1

    #claculate most frequent symbol and it's count
    def calculate(self):
        self.symbol,self.count = max(self.length.items(),
        key=operator.itemgetter(1))

    #verify if wrong symbol is subtracted ie ( V, ..
    def verifySubstract(self, current):
        while current % 10 != 0:
            current = current / 10
```

```python
        if current == 5:
            raise Exception("Wrong Substraction")

    def evaluate(self):
        #check mximum repeatation is crossing the limit
        if self.count > 3:
            raise Exception("Repeat more than 3")
        #symbol is proper or not
        if self.symbol not in self.compiler.symbol_map:
            raise Exception("Wrong Symbol")
        #check if wrong symbol is repeated ie (V, ..
        self.symbol,unit = self.compiler.
        evaluateSymbol(self.symbol)
        while self.symbol % 10 != 0:
            self.symbol = self.symbol / 10
        if self.count > 1 and  self.symbol == 5:
            raise Exception("Wrong Symbol repeated")

    #checking if input sentence is proper or not
    def evaluateSentence(self, line):
        if "is" not in line:
            return "I have no idea what you are
            talking about"
```

The compilation logic is in the compiler class, as shown here:

```python
import sys

from observer import Observer

class compilerTrader(object):
    #store mapping of symbols with score and unit
    symbol_map = {}
    #store the list of valid symbol
    valid_values = []
```

```python
#read the config and initialize the class member
def __init__(self, config_path):
    with open(config_path) as f:
        for line in f:
            if ',' in line :
                symbol, value, type = line.strip().
                split(',')
                self.symbol_map[symbol] = float(value), type
                self.valid_values.append(float(value))
        f.close()

#evaluate the ultimate numerical score with unit for
a symbol
def evaluateSymbol(self, symbol):
    while symbol not in self.valid_values:
            symbol,unit = self.symbol_map[symbol]
    return float(symbol), unit

#compiling the info in line
def compile_super(self, line):
    obs = Observer(self)
    if 'is' in line:
        fields = line.split(' is ')
        value = fields[-1]
        var = fields[0]
        #if one symbol and one value
        if ' ' not in var:
            if value in self.symbol_map:
                self.symbol_map[var] = int(self.symbol_
                map[value][0]) ,'roman'
```

```python
        else:
            #logic for value with unit
            if ' ' in value:
                fields = value.split(' ')
                user_unit = fields[-1]
                if ' ' not in var:
                    self.symbol_map[var] =
                    int(fields[0]), user_unit
                else:
                    #logic for multiple symbols
                    in input
                    total = int(fields[0])
                    factor = 0
                    arr = var.split(' ')
                    obs.initialize(arr)
                    for i in range(len(arr)):
                        obs.increment(arr[i])
                        if arr[i] in self.symbol_map
                        and arr[i+1] in self.symbol_map
                        and i < len(arr) -1:
                            current, current_
                            unit = self.
                            evaluateSymbol(
                            [arr[i]][0])
                            next, next_unit = self.
                            evaluateSymbol(
                            [arr[i+1]][0])
                            if current >= next:
                                factor =  factor
                                + current
```

```
        else:
            obs.verifySubstract(
            current)
            factor = factor -
            current
    else:
        if arr[i] in self.
        symbol_map:
        current, current_
        unit = self.
        evaluateSymbol(
        [arr[i]][0])
        factor = factor +
        current
            else:
                self.
                symbol_map[
                arr[i]] =
                total/factor,
                user_unit
                self.valid_
                values.append
                (total/factor)
    obs.calculate()
    obs.evaluate()
```

The answering logic is in the answer layer, which calls Observer and compiler. The answering class inherits the compiler class.

```
import sys

from observer import Observer
from compiler import compilerTrader
```

```python
class answeringTrader(compilerTrader):

    def __init__(self, config_path):
        super().__init__(config_path)

    #compiling info in line
    def compile(self, line):
        super().compile_super(line)

    #answering query in line
    def answer(self, line):
        obs = Observer(super())
        if 'is' in line:
            values = line.split(' is ')[-1]
            ans = 0
            arr = values.split(' ')
            unit = ''
            obs.initialize(arr)
            for i in range(len(arr)):
                if arr[i] in "?.,!;":
                    continue
                obs.increment(arr[i])
                if i < len(arr)-2:
                    if arr[i] in super().symbol_
                    map and arr[i+1] in super().
                    symbol_map:
                        current, current_
                        unit = super().
                        evaluateSymbol([arr[i]][0])
                        next, next_unit = super().
                        evaluateSymbol(
                        [arr[i+1]][0])
```

```
                    if current >= next:
                        ans = ans + current
                    else:
                        if next_unit ==
                        'roman':
                            obs.verify
                            Substract(
                            current)
                            ans = ans -
                            current
                        else:
                            ans = ans
                            + current
            else:
                if arr[i] in super().symbol_map:
                    current,unit = super().
                    evaluateSymbol([arr[i]][0])
                    if  unit != 'roman':
                        ans = ans * current
                    else:
                        ans = ans + current

    obs.calculate()
    obs.evaluate()
    values = values.replace("?" , "is ")
    if unit == 'roman':
        unit = ''
    return(values + str(ans) + ' ' + unit)
```

Finally, the main program calls the answering class and the observer, and then it performs the task and does unit testing on the logic.

```python
import sys
import unittest

sys.path.append('./answerLayer')
sys.path.append('./compilerLayer')
sys.path.append('./utilityLayer')

from answer import answeringTrader
from observer import Observer

#client interface for the framework
class ClientTrader(object):
    trader = None
    def __init__(self, config_path):
        self.trader = answeringTrader(config_path)

    #processing an input string
    def process(self, input_string):
        obs = Observer(self.trader)
        valid = obs.evaluateSentence(input_string)
        if valid is not None:
            return valid
        if input_string.strip()[-1] == '?' :
            return self.trader.answer(input_string)
        else:
            return self.trader.compile(input_string)

#unit test cases
class TestTrader(unittest.TestCase):
    trader = None

    def setUp(self):
        pass
```

```python
#test case for non-roman symbol       unit other
than roman
def test_answer_unit(self):
    ans = self.trader.process("how many Credits is glob
    prok Silver ?")
    self.assertEqual(ans.strip(), "glob prok Silver is
    68.0 Credits")

#test case with only roman symbol in unit case
def test_answer_roman(self):
    ans = self.trader.process("how much is pish tegj
    glob glob ?")
    self.assertEqual(ans.strip(), "pish tegj glob glob
    is 42.0")

#test case if repeatation of symbol is exceed max
limit (3)
def test_exception_over_repeat(self):
    with self.assertRaises(Exception) as context:
        ans = self.trader.process("how much is pish
        tegj glob glob glob glob ?")
        self.assertTrue("Repeat more than 3" in
        context.exception)

#test case if wrong symbol repeated ie (V ..
def test_exception_unproper_repeat(self):
    with self.assertRaises(Exception) as context:
        ans = self.trader.process("how much is pish
        tegj D D ?")
        self.assertTrue("Repeat more than 3" in
        context.exception)
```

```python
        #test case if wrong symbol substracted ie (V ...
        def test_wrong_substraction(self):
                with self.assertRaises(Exception) as context:
                        ans = self.trader.process("how much
                        is V X ?")
                        self.assertTrue("Wrong Substraction" in
                        context.exception)

        #test case if query is not properly formatted
        def test_wrong_format_query(self):
                ans = self.trader.process("how much wood could a
                woodchuck chuck if a woodchuck could chuck wood ?")
                self.assertEqual(ans.strip(), "I have no idea what
                you are talking about")

if __name__ == '__main__':
    if len(sys.argv) != 3:
            print("Usage is : " + sys.argv[0] + " <intput file
            path>  <config file path>")
            exit(0)
    tr = ClientTrader(sys.argv[2])
    f = open(sys.argv[1])
    for line in f:
            response = tr.process(line.strip())
            if response is not None:
                    print(response)

    TestTrader.trader = tr
    unittest.main(argv = [sys.argv[0]], exit = False)
```

You can run this program with the following command:

```
python client.py input.txt config.txt
```

Calling Other Languages in Python

Now we will describe how to use other languages in Python. There are two examples here. The first is calling R code from Python. R code is required for some use cases. For example, if you want a ready-made function for the Holt-Winters method in a time series, it is difficult to perform in Python, but it is available in R. So, you can call R code from Python using the rpy2 module, as shown here:

```
import rpy2.robjects as ro
ro.r('data(input)')
ro.r('x <-HoltWinters(input data frame)')
```

(You can use example data given in time series chapter.)

Sometimes you need to call Java code from Python. For example, say you are working on a name-entity recognition problem in the field of natural language processing (NLP); some text is given as input, and you have to recognize the names in the text. Python's NLTK package does have a name-entity recognition function, but its accuracy is not good. Stanford NLP is a better choice here, but it is written in Java. You can solve this problem in two ways.

- You can call Java at the command line using Python code. You need to install Java with the yum/at-get install java command before calling it.

- For Windows, it is recommended that you install the JRE from https://adoptium.net/temurin/releases/?version=8. You can also install the JRE from another distribution. The installation will automatically create JAVA_HOME. If it does not, you need to set JAVA_HOME as the system variable, and the value should be the location of Java installation folder, for example, JAVA_HOME=C:\Program Files\Eclipse Adoptium\jdk-8.0.345.1-hotspot\.

```
import subprocess
subprocess.call(['java','-cp','*','edu.stanford.nlp.sentiment.
SentimentPipeline','-file','foo.txt'])
```

Please place foo.txt in the same folder where you run the Python code.

- You can expose Stanford NLP as a web service and call it as a service. (Before running this code, you'll need to download the Stanford nlp JAR file available with the book's source code.)

```
nlp = StanfordCoreNLP('http://127.0.0.1:9000')
output = nlp.annotate(sentence, properties={
 "annotators": "tokenize,ssplit,parse,sentiment",
 "outputFormat": "json",
 # Only split the sentence at End Of Line. We assume that this
 method only takes in one single sentence.
 "ssplit.eolonly": "true",
 # Setting enforceRequirements to skip some annotators and make
 the process faster
 "enforceRequirements": "false"
 })
```

You will see a more detailed example of Stanford NLP in Chapter 2.

Exposing the Python Model as a Microservice

You can expose the Python model as a microservice in the same way that your Python model can be used by others to write their own code. The best way to do this is to expose your model as a web service. As an example, the following code exposes a deep learning model using Flask:

```python
from flask import Flask, request, g
from flask_cors import CORS
import tensorflow as tf
from sqlalchemy import *
from sqlalchemy.orm import sessionmaker
import pygeoip
from pymongo import MongoClient
import json
import datetime as dt
import ipaddress
import math
app = Flask(__name__)
CORS(app)
@app.before_request
def before():
      db = create_engine('sqlite:///score.db')
      metadata = MetaData(db)
      g.scores = Table('scores', metadata, autoload=True)
      Session = sessionmaker(bind=db)
      g.session = Session()
      client = MongoClient()
      g.db = client.frequency
      g.gi = pygeoip.GeoIP('GeoIP.dat')
      sess = tf.Session()
      new_saver = tf.train.import_meta_graph('model.obj.meta')
      new_saver.restore(sess, tf.train.latest_
      checkpoint('./'))
      all_vars = tf.get_collection('vars')
      g.dropped_features = str(sess.run(all_vars[0]))
      g.b = sess.run(all_vars[1])[0]
      return
```

```python
def get_hour(timestamp):
    return dt.datetime.utcfromtimestamp(timestamp /
    1e3).hour
def get_value(session, scores, feature_name, feature_value):
    s = scores.select((scores.c.feature_name == feature_
    name) & (scores.c.feature_value == feature_value))
    rs = s.execute()
    row = rs.fetchone()
    if row is not None:
        return float(row['score'])
    else:
        return 0.0
@app.route('/predict', methods=['POST'])
def predict():
    input_json = request.get_json(force=True)
    features = ['size','domain','client_time','device','ad_
    position','client_size', 'ip','root']
    predicted = 0
    feature_value = ''
    for f in features:
        if f not in g.dropped_features:
            if f == 'ip':
                feature_value = str(ipaddress.
                IPv4Address(ipaddress.ip_
                address(unicode(request.remote_
                addr))))
            else:
                feature_value = input_json.get(f)
            if f == 'ip':
                if 'geo' not in g.dropped_features:
```

```
                    geo = g.gi.country_name_by_
                    addr(feature_value)
                    predicted = predicted + get_
                    value(g.session, g.scores,
                    'geo', geo)
        return str(math.exp(predicted + g.b)-1)
app.run(debug = True, host ='0.0.0.0')
```

This code exposes a deep learning model as a Flask web service. A JavaScript client will send the request with web user parameters such as the IP address, ad size, ad position, and so on, and it will return the price of the ad as a response. The features are categorical. You will learn how to convert them into numerical scores in Chapter 3. These scores are stored in an in-memory database. The service fetches the score from the database, sums the result, and replies to the client. This score will be updated real time in each iteration of training of a deep learning model. It is using MongoDB to store the frequency of that IP address in that site. It is an important parameter because a user coming to a site for the first time is really searching for something, which is not true for a user where the frequency is greater than 5. The number of IP addresses is huge, so they are stored in a distributed MongoDB database.

High-Performance API and Concurrent Programming

Flask is a good choice when you are building a general solution that is also a graphical user interface (GUI). But if high performance is the most critical requirement of your application, then Falcon is the best choice. The following code is an example of the same model shown previously exposed by the Falcon framework. Another improvement we made in this code is that we implemented multithreading, so the code will be executed in parallel. In addition to the Falcon-specific changes, you should note the major changes in parallelizing the calling get_score function using a thread pool class.

```python
import falcon
from falcon_cors import CORS
import json
from sqlalchemy import *
from sqlalchemy.orm import sessionmaker
import pygeoip
from pymongo import MongoClient
import json
import datetime as dt
import ipaddress
import math
from concurrent.futures import *
from sqlalchemy.engine import Engine
from sqlalchemy import event
import sqlite3
@event.listens_for(Engine, "connect")
def set_sqlite_pragma(dbapi_connection, connection_record):
 cursor = dbapi_connection.cursor()
 cursor.execute("PRAGMA cache_size=100000")
 cursor.close()
class Predictor(object):
    def __init__(self,domain):
        db1 = create_engine('sqlite:///score_' + domain +
        '0test.db')
        metadata1 = MetaData(db1)
        self.scores = Table('scores', metadata1,
        autoload=True)
        client = MongoClient(connect=False,maxPoolSize=1)
```

```python
        self.db = client.frequency
        self.gi = pygeoip.GeoIP('GeoIP.dat')
        self.high = 1.2
        self.low = .8
def get_hour(self,timestamp):
        return dt.datetime.utcfromtimestamp(timestamp /
        1e3).hour
def get_score(self, featurename, featurevalue):
        pred = 0
        s = self.scores.select((self.scores.c.feature_name
        == featurename) & (self.scores.c.feature_value ==
        featurevalue))
        rs = s.execute()
        row = rs.fetchone()
        if row is not None:
                pred = pred + float(row['score'])
        res = self.db.frequency.find_one({"ip" : ip})
        freq = 1
        if res is not None:
                freq = res['frequency']
                pred2, prob2 = self.get_score('frequency',
                str(freq))
                return (pred1 + pred2), (prob1 + prob2)

    conn = sqlite3.connect('multiplier.db')
                cursor = conn.execute("SELECT high,low from
                multiplier where domain='" + value + "'")
                row = cursor.fetchone()
                if row is not None:
                        self.high = row[0]
                        self.low = row[1]
        return self.get_score(f, value)
```

```python
    def on_post(self, req, resp):
        input_json = json.loads(req.stream.
        read(),encoding='utf-8')
        input_json['ip'] = unicode(req.remote_addr)
        pred = 1
        prob = 1
        with ThreadPoolExecutor(max_workers=8) as pool:
            future_array = { pool.submit(self.
            get_value,f,input_json[f]) : f for f in
            input_json}
            for future in as_completed(future_array):
                pred1, prob1 = future.result()
                pred = pred + pred1
                prob = prob - prob1
        resp.status = falcon.HTTP_200
res = math.exp(pred)-1
        if res < 0:
            res = 0
        prob = math.exp(prob)
        if(prob <= .1):
            prob = .1
        if(prob >= .9):
            prob = .9
        multiplier = self.low + (self.high -self.low)*prob
        pred = multiplier*pred
        resp.body = str(pred)
cors = CORS(allow_all_origins=True,allow_all_
methods=True,allow_all_headers=True)
wsgi_app = api = falcon.API(middleware=[cors.middleware])
f = open('publishers1.list')
```

```
for domain in f:
    domain = domain.strip()
    p = Predictor(domain)
    url = '/predict/' + domain
    api.add_route(url, p)
```

Having covered design patterns in Python a bit, let's now take a look at some essential architecture patterns for data scientists.

Choosing the Right Database

Before we go, we'll leave a note for manager on which database is best for which case.

- A relational database (MySQL, Oracle, SQL Server) is the preferable choice when data is highly structured and entities have a clear and strict connection. Mongo, on the other hand, is a better choice when data is unstructured and unorganized.

- Elastic Search or Solr is a better choice when data contains a lengthy textual field and you're executing lots of searches in a substring of the text field. With Elastic Search, you get a free data visualization tool called Kibana as well as an ETL tool called Logstash, and full-stack data analytics solutions are fashionable.

- Data must sometimes be represented as a graph. In that situation, a graph database is required. Neo4j is a popular graph database that comes with many utility tools at a low price.

- We occasionally require a quick application. In that situation, an in-memory database like SQLite can be used. However, SQLite does not support updating your database from a remote host.

You'll learn more about databases in Chapter 2.

Summary

In this chapter, we discussed fundamental engineering principles for data scientists, which are covered in separate chapters. The question-answering example can help you understand how to organize your code. The basic rule is to not put everything into one class. Divide your code into many categories and use parent-child relationships where they exist. Then you learned how to use Python to call other languages' code. We provided two instances of R and Java code calls. Then we showed you how to expose your model as a REST API and make it perform well by using concurrent programming. Following that, we covered significant architectural patterns from data scientists.

CHAPTER 2

ETL with Python

Every data science professional has to extract, transform, and load (ETL) data from different data sources. In this chapter, we will discuss how to perform ETL with Python for a selection of popular databases. For a relational database, we'll cover MySQL. As an example of a document database, we will cover Elasticsearch. For a graph database, we'll cover Neo4j, and for NoSQL, we'll cover MongoDB. We will also discuss the Pandas framework, which was inspired by R's data frame concept.

ETL is based on a process in which data is extracted from multiple sources, transformed into specific formats that involve cleaning enrichment, and finally loaded into its target destination. The following are the details of each process:

1. *Extract*: During data extraction, source data is pulled from a variety of sources and moved to a staging area, making the data available to subsequent stages in the ETL process. After that, the data undergoes the cleaning and enrichment stage, also known as data cleansing.

2. *Transform*: In this stage, the source data is matched to the format of the target system. This includes steps such as changing data types, combining fields, splitting fields, etc.

© Sayan Mukhopadhyay, Pratip Samanta 2023
S. Mukhopadhyay and P. Samanta, *Advanced Data Analytics Using Python*,
https://doi.org/10.1007/978-1-4842-8005-8_2

3. *Load*: This stage is the final ETL stage. Here, data is loaded into the data warehouse in an automated manner and can be periodically updated. Once completed, the data is ready for data analysis.

The previous processes are important in any data analytics work. Once the data goes through the ETL processes, then it becomes possible to analysis the data, find insights, and so on.

We will discuss various types of ETL throughout this chapter. We discussed in Chapter 1 that data is not an isolated thing. We need to load data from somewhere, which is a database. We need to fetch the data from some application, which is extraction. In this chapter and the next, we will discuss various feature engineering that transforms the data from one form to another.

MySQL

MySQLdb is an API in Python developed to work on top of the MySQL C interface.

How to Install MySQLdb?

First you need to install the Python MySQLdb module on your machine. Then run the following script:

```
#!/usr/bin/python
import MySQLdb
```

If you get an import error exception, that means the module was not installed properly.

The following are the instructions to install the MySQL Python module:

```
$ gunzip MySQL-python-1.2.2.tar.gz
```

```
$ tar -xvf MySQL-python-1.2.2.tar
$ cd MySQL-python-1.2.2
$ python setup.py build
$ python setup.py install
```

You can download the `tar.gz` file from `https://dev.mysql.com/downloads/connector/python/`. You need to download it to your working folder.

For Windows, please select the MySQL installer file from `https://dev.mysql.com/downloads/installer/`. Once it's downloaded, double-click the file to install it and select MySQL Connector/Python as one of the products to install. For details, you can visit `https://dev.mysql.com/doc/connector-python/en/connector-python-installation-binary.html`.

Database Connection

Before connecting to a MySQL database, make sure you do the following:

1. You need to access a database called TEST with the `sql "use test"` command.

2. In TEST you need a table named STUDENT; use the command `sql "create table student(name varchar(20), sur_name varchar(20),roll_ no int")`;.

3. STUDENT needs three fields: NAME, SUR_NAME, and ROLL_NO.

4. There needs to be a user in TEST that has complete access to the database.

If you do not do these steps properly, you will get an exception in the next Python code.

INSERT Operation

The following code carries out the SQL INSERT statement for the purpose of creating a record in the STUDENT table:

```
#!/usr/bin/python
import MySQLdb
# Open database connection
db = MySQLdb.connect("localhost","user","passwd","TEST" )
# prepare a cursor object using cursor() method
cursor = db.cursor()
# Prepare SQL query to INSERT a record into the database.
sql = """INSERT INTO STUDENT(NAME,
        SUR_NAME, ROLL_NO)
        VALUES ('Sayan', 'Mukhopadhyay', 1)"""
try:
   # Execute the SQL command
   cursor.execute(sql)
   # Commit your changes in the database
   db.commit()
except:
   # Rollback in case there is any error
   db.rollback()
# disconnect from server
db.close()
```

READ Operation

The following code fetches data from the STUDENT table and prints it:

```
#!/usr/bin/python
import MySQLdb
# Prepare SQL query to INSERT a record into the database.
```

```
sql = "SELECT * FROM STUDENT "
try:
   # Execute the SQL command
   cursor.execute(sql)
   # Fetch all the rows in a list of lists.
   results = cursor.fetchall()
   for row in results:
       fname = row[0]
       lname = row[1]
       id = row[2]
     # Now print fetched result
Print( "name=%s,surname=%s,id=%d" % \
          (fname, lname, id ))
except:
    print "Error: unable to fecth data"
```

DELETE Operation

The following code deletes a row from TEST with id=1:

```
#!/usr/bin/python
import MySQLdb
# Prepare SQL query to DELETE required records
sql = "DELETE FROM STUDENT WHERE ROLL_NO =1"
try:
   # Execute the SQL command
   cursor.execute(sql)
   # Commit your changes in the database
   db.commit()
except:
   # Rollback in case there is any error
   db.rollback()
```

UPDATE Operation

The following code changes the lastname variable to Mukherjee, from Mukhopadhyay:

```
#!/usr/bin/python
import MySQLdb
# Prepare SQL query to UPDATE required records
sql = "UPDATE STUDENT SET SUR_NAME="Mukherjee"
                        WHERE SUR_NAME="Mukhopadhyay"
try:
   # Execute the SQL command
   cursor.execute(sql)
   # Commit your changes in the database
   db.commit()
except:
   # Rollback in case there is any error
   db.rollback()
```

COMMIT Operation

The commit operation provides its assent to the database to finalize the modifications, and after this operation, there is no way that this can be reverted.

ROLL-BACK Operation

If you are not completely convinced about any of the modifications and you want to reverse them, then you can apply the roll-back() method.

The following is a complete example of accessing MySQL data through Python. It will give the complete description of the data stored in a CSV file or MySQL database.

This code asks for the data source type, either MySQL or text. For example, if MySQL asks for the IP address, credentials, and database name and shows all tables in the database, it offers its fields once the table is selected. Similarly, a text file asks for a path, and in the files it points to, all the columns are shown to the user.

```python
# importing files and reading config file

import MySQLdb
import sys
out = open('Config1.txt','w')
print ("Enter the Data Source Type:")
print( "1. MySql")
print ("2. Exit")
while(1):
        data1 = sys.stdin.readline().strip()
        if(int(data1) == 1):
                out.write("source begin"+"\n"+"type=mysql\n")

# taking inputs from user

                print ("Enter the ip:")
                ip = sys.stdin.readline().strip()
                out.write("host=" + ip + "\n")
                print ("Enter the database name:")
                db = sys.stdin.readline().strip()
                out.write("database=" + db + "\n")
                print ("Enter the user name:")
                usr = sys.stdin.readline().strip()
                out.write("user=" + usr + "\n")
                print ("Enter the password:")
                passwd = sys.stdin.readline().strip()
                out.write("password=" + passwd + "\n")
```

making connection to and executing query

```
            connection = MySQLdb.connect(ip, usr, passwd, db)
            cursor = connection.cursor()
            query = ("show tables")
            cursor.execute(query)
            data = cursor.fetchall()
            tables = []
```

appending data to the table

```
            for row in data:
                    for field in row:
                            tables.append(field.strip())
            for i in range(len(tables)):
                    print( i, tables[i])
            tb = tables[int(sys.stdin.readline().strip())]
            out.write("table=" + tb + "\n")
            query = ("describe " + tb)
            cursor.execute(query)
            data = cursor.fetchall()
            columns = []
            for row in data:
                    columns.append(row[0].strip())
            for i in range(len(columns)):
                    print( columns[i])
            print "Not index choose the exact column names
            seperated by coma"
            cols = sys.stdin.readline().strip()
            out.write("columns=" + cols + "\n")
            cursor.close()
            connection.close()
            out.write("source end"+"\n")
```

```
print ("Enter the Data Source Type:")
print ("1. MySql")              print ("2. Exit")
out.close()
sys.exit()
```

Before we go on to the topic of relational databases, let's talk about database normalization.

Normal Forms

Database normal forms are the principles to organize your data in an optimum way.

Every table in a database can be in one of the normal forms that we'll go over next. For the primary key (PK) and foreign key (FK), you want to have as little repetition as possible. The rest of the information should be taken from other tables.

- First normal form (1NF)

- Second normal form (2NF)

- Third normal form (3NF)

First Normal Form

There are no repeating groups in the *first normal form* since only single values are allowed at the intersection of each row and column.

To get to first normal form, remove the repetitive groups, and establish two new relations to normalize a connection with a repeating group.

For unique identification, the new connection's PK is a combination of the old relation's PK and a feature from the newly formed relation.

To demonstrate the procedure for 1NF, we'll use the `Student_Grade_Report` table, which comes from a `School` database.

Student_Grade_Report (StudentNo, StudentName, Major, CourseNo, CourseName, InstructorNo, InstructorName, InstructorLocation, Grade)

1. The recurring group in the Student Grade Report table contains the course information. A student can enroll in a variety of courses.

2. Get rid of the group that keeps repeating itself. That's each student's course information in this situation.

3. Determine your new table's PK.

4. The attribute value must be identified uniquely by the PK (StudentNo and CourseNo).

Student (StudentNo, StudentName, Major)
StudentCourse (StudentNo, CourseNo, CourseName, InstructorNo, InstructorName, InstructorLocation, Grade)

Second Normal Form

The relation must first be in 1NF for the *second normal form*. If and only if the PK contains a single feature, the relationship is automatically in 2NF.

If the connection contains a composite PK, then each nonkey property must be completely reliant on the entire PK, not just a portion of it (i.e., there can't be any partial augmentation or dependency).

A table must first be in 1NF before moving to 2NF.

1. As it has a single-column PK, the Student table is already in 2NF.

2. When looking at the Student Course table, you can observe that not all of the characteristics, especially the course details, are completely dependent on the PK. The grade is the sole attribute that is entirely reliant on xxx.

3. Locate the new table containing the course details.

4. Determine the new table's PK.

The three new tables are as follows:

Student (StudentNo, StudentName, Major)
CourseGrade (StudentNo, CourseNo, Grade)
CourseInstructor (CourseNo, CourseName, InstructorNo, InstructorName, InstructorLocation)

Third Normal Form

The connection must be in second normal form to be in *third normal form*. All transitive dependencies must be eliminated as well; a nonkey attribute cannot be functionally reliant on another nonkey attribute.

This is the process for achieving 3NF:

1. From each table with a transitive relationship, remove all dependent characteristics in a transitive relationship.

2. Make a new table with the dependence eliminated.

3. Inspect new and updated tables to ensure that each table has a determinant and that no tables have improper dependencies.

Take a look at the four new tables:

Student (StudentNo, StudentName, Major)
CourseGrade (StudentNo, CourseNo, Grade)
Course (CourseNo, CourseName, InstructorNo)
Instructor (InstructorNo, InstructorName, InstructorLocation)

There should be no abnormalities in the third normal form at this point. For this example, consider the dependency diagram in Figure 2-1. As previously said, the first step is to eliminate repeated groupings.

Student (StudentNo, StudentName, Major)
StudentCourse (StudentNo, CourseNo, CourseName, InstructorNo, InstructorName, InstructorLocation, Grade)

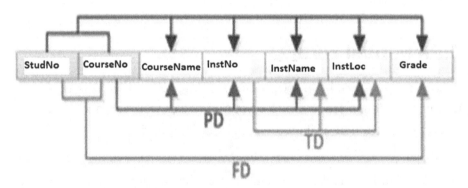

Figure 2-1. *Dependency diagram*

Review the dependencies in Figure 2-1, which summarizes the normalization procedure for the School database.

The following are the abbreviations used in Figure 2-1:

- PD stands for partially dependent.

- TD stands for transitive dependence.

- FD stands for full dependency. (FD stands for functional dependence in most cases. Figure 2-1 is the only place where FD is used as an abbreviated form for full dependence.)

A relational database is valuable when structured data and a strict relationship between the fields are maintained. But what if you do not have structured data in which a strict relationship between fields has been maintained? That's where Elasticsearch comes in.

Elasticsearch

You'll find that data is often unstructured. Meaning, you may end up with a mix of image data, sensor data, and other forms of data. To analyze this data, we first need to store it. MySQL or SQL-based databases are not good at storing unstructured data. So here we introduce a different kind of storage, which is mainly used to handle unstructured textual data.

Elasticsearch is a Lucene-based database, which makes it is easy to store and search text data. Its query interface is a REST API endpoint. The Elasticsearch (ES) low-level client gives a direct mapping from Python to ES REST endpoints. One of the big advantages of Elasticsearch is that it provides a full-stack solution for data analysis in one place. Elasticsearch is the database. It has a configurable front end called Kibana, a data collection tool called Logstash, and an enterprise security feature called Shield.

This example has features called cat, cluster, indices, ingest, nodes, snapshot, and tasks that translate to instances of `CatClient`, `ClusterClient`, `IndicesClient`, `CatClient`, `ClusterClient`, `IndicesClient`, `IngestClient`, `NodesClient`, `SnapshotClient`, `NodesClient`, `SnapshotClient`, and `TasksClient`, respectively. These instances are the only supported way to get access to these classes and their methods.

You can specify your own connection class, which can be used by providing the connection_class parameter.

```
# create connection to local host using the ThriftConnection
Es1=Elasticsearch(connection_class=ThriftConnection)
```

Installation commands for Elastic Search are given here:

```
curl -fsSL https://artifacts.elastic.co/GPG-KEY-elasticsearch |
sudo apt-key add -
echo "deb https://artifacts.elastic.co/packages/7.x/apt stable
main" | sudo tee -a /etc/apt/sources.list.d/elastic-7.x.list
sudo apt update
sudo apt install elasticsearch
```

You can start Elasticsearch in Ubuntu with these commands:

```
service elasticsearch start
service elasticsearch stop
```

You can check the status with these commands:

```
service elasticsearch status
```

```
# create connection that will automatically inspect the
cluster to get
# the list of active nodes. Start with nodes running on
'esnode1' and
# 'esnode2'
Es1=Elasticsearch(
    ['esnode1', 'esnode2'],
# sniff before doing anything
sniff_on_start=True,
# refresh nodes after a node fails to respond
sniff_on_connection_fail=True,
```

```
# and also every 30 seconds
sniffer_timeout=30
)
```

Different hosts can have different parameters (hostname, port number, SSL option); you can use one dictionary per node to specify them.

```
# connect to localhost directly and
another node using SSL on port 443
# and an url_prefix. Note that ``port`` needs to be an int.
Es1=Elasticsearch([
{'host':'localhost'},
{'host':'othernode','port':443,'url_prefix':'es','use_
ssl':True},
])
```

SSL client authentication is also supported (see Urllib3HttpConnection for a detailed description of the options); an example is given here:

```
Es1=Elasticsearch(
['localhost:443','other_host:443'],
# turn on SSL
use_ssl=True,
# make sure we verify SSL certificates (off by default)
verify_certs=True,
# provide a path to CA certs on disk
ca_certs='path to CA_certs',
# PEM formatted SSL client certificate
client_cert='path to clientcert.pem',
# PEM formatted SSL client key
client_key='path to clientkey.pem'
)
```

Connection Layer API

Many classes are responsible for dealing with the Elasticsearch cluster. Here, the default subclasses being utilized can be disregarded by handing over parameters to the Elasticsearch class. Every argument belonging to the client will be added onto Transport, ConnectionPool, and Connection.

As an example, if you want to use your own personal utilization of the ConnectionSelector class, you just need to pass in the selector_class parameter.

The entire API wraps the raw REST API with a high level of accuracy, which includes the differentiation between the required and optional arguments to the calls. This implies that the code makes a differentiation between positional and keyword arguments; I advise you to use keyword arguments for all calls to be consistent and safe. An API call becomes successful (and will return a response) if Elasticsearch returns a 2XX response. Otherwise, an instance of TransportError (or a more specific subclass) will be raised. You can see other exceptions and error states in exceptions. If you do not want an exception to be raised, you can always pass in an ignore parameter with either a single status code that should be ignored or a list of them.

```
from elasticsearch import Elasticsearch
es=Elasticsearch()
# ignore 400 cause by IndexAlreadyExistsException when creating
an index
es.indices.create(index='test-index',ignore=400)
# ignore 404 and 400
es.indices.delete(index='test-index',ignore=[400,404])
```

Neo4j Python Driver

There are a variety of systems, such as network topology and social networks. However, when difficulties are shown as a graph, they are quickly resolved. Neo4j is a database that stores data in the form of a graph and executes queries through a graphical interface. The Neo4j Python driver is supported by Neo4j and connects with the database through the binary protocol. It tries to remain minimalistic but at the same time be idiomatic to Python.

```
pip install neo4j-driver
from neo4j.v1 import GraphDatabase, basic_auth
driver11 = GraphDatabase.driver("bolt://localhost", auth=basic_
auth("neo4j", "neo4j"))
session11 = driver11.session()
session11.run("CREATE (a:Person {name:'Sayan',
title:'Mukhopadhyay'})")
result 11= session11.run("MATCH (a:Person) WHERE a.name =
'Sayan' RETURN a.name AS name, a.title AS title")
for recordi n resul11t:
print("%s %s"% (record["title"], record["name"]))
session.close()
```

neo4j-rest-client

The main objective of neo4j-rest-client is to make sure that the Python programmers already using Neo4j locally through python-embedded are also able to access the Neo4j REST server. So, the structure of the neo4j-rest-client API is completely in sync with python-embedded. But, a new structure is brought in so as to arrive at a more Pythonic style and to augment the API with the new features being introduced by the Neo4j team.

In-Memory Database

Another important class of databases is an in-memory database. This type stores and processes the data in RAM. So, operations on the database are fast, and the data is volatile. SQLite is a popular example of an in-memory database. In Python you need to use the sqlalchemy library to operate on SQLite. In Chapter 1's Flask and Falcon example, I showed you how to select data from SQLite. Here I will show how to store a Pandas data frame in SQLite:

```
from sqlalchemy import create_engine
import sqlite3
conn = sqlite3.connect('multiplier.db')
conn.execute('''CREATE TABLE if not exists multiplier
        (domain          CHAR(50),
          low          REAL,
          high          REAL);''')
conn.close()
db_name = "your db name ""
disk_engine = create_engine(db_name)
df.to_sql('scores', disk_engine, if_exists='replace')
```

MongoDB (Python Edition)

MongoDB is an open-source *document database* designed for superior performance, easy availability, and automatic scaling. MongoDB makes sure that object-relational mapping (ORM) is not required to facilitate development. A document that contains a data structure made up of field and value pairs is referred to as a *record* in MongoDB. These records are akin to JSON objects. The values of fields may be comprised of other documents, arrays, and arrays of documents.

```
{
"_id":ObjectId("01"),
"address": {
"street":"Siraj Mondal Lane",
"pincode":"743145",
"building":"129",
"coord": [ -24.97, 48.68 ]
    },
"borough":"Manhattan",
```

Import Data into the Collection

mongoimport can be used to place the documents into a collection in a database, within the system shell or a command prompt. If the collection already exists in the database, the operation will discard the original collection first.

```
mongoimport --DB test --collection restaurants --drop --file ~/
downloads/primer-dataset.json
```

The mongoimport command is joined to a MongoDB instance running on localhost on port 27017. The --file option provides a way to import the data; here it's ~/downloads/primer-dataset.json.

To import data into a MongoDB instance running on a different host or port, the hostname or port needs to be mentioned specifically in the mongoimport command by including the --host or --port option.

There is a similar load command in MySQL.

Create a Connection Using pymongo

To create a connection, do the following:

```
import MongoClient from pymongo.
Client11 = MongoClient()
```

If no argument is mentioned to MongoClient, then it will default to the MongoDB instance running on the localhost interface on port 27017.

A complete MongoDB URL may be designated to define the connection, which includes the host and port number. Let's take a look at an example.

First, install Mongo using this command: yum/apt install mongo.

Then, launch MongoDB using this command: service mongo start.

The following code makes a connection to a MongoDB instance that runs on mongodb0.example.net and port 27017:

```
Client11 = MongoClient("mongodb://myhostname:27017")
```

Access Database Objects

To assign the database named primer to the local variable DB, you can use either of the following lines:

```
Db11 = client11.primer
db11 = client11['primer']
```

Collection objects can be accessed directly by using the dictionary style or the attribute access from a database object, as shown in the following two examples:

```
Coll11 = db11.dataset
coll = db11['dataset']
```

Insert Data

You can place a document into a collection that doesn't exist, and the following operation will create the collection:

```
result=db.addrss.insert_one({<<your json >>)
```

Update Data

Here is how to update data:

```
result=db.address.update_one(
 {"building": "129",
 {"$set": {"address.street": "MG Road"}}
)
```

Remove Data

To expunge all documents from a collection, use this:

```
result=db.restaurants.delete_many({})
```

Cloud Databases

Even though the cloud has its own chapter, we'd like to provide you with an overview of cloud databases, particularly databases for large data. People prefer cloud databases when they want their systems to scale automatically. Google Big Query is the greatest tool for searching your data. Azure Synapsys has a similar feature; however, it is significantly more expensive. You can store data on S3, but if you want to run a query, you'll need Athena, which is expensive. So, in modern practice, data is stored as a blob in S3, and everything is done in a Python application. If there is an error in data finding, this method takes a long time. Amazon Redish can also handle a considerable quantity of large data and comes with a built-in BI tool.

Pandas

The goal of this section is to show some examples to enable you to begin using Pandas. These illustrations have been taken from real-world data, along with any bugs and weirdness that are inherent. Pandas is a framework inspired by the R data frame concept.

Please find the CSV file at the following link:

https://github.com/Apress/advanced-data-analytics-python-2e

To read data from a CSV file, use this:

```
import pandas as pd
broken_df=pd.read_csv('fetaure_engineering_data.csv')
```

To look at the first three rows, use this:

```
broken_df[:3]
```

To select a column, use this:

```
broken_df[' MSSubClass ']
```

To plot a column, use this:

```
broken_df[' MSSubClass' '].plot()
```

To get a maximum value in the data set, use this:

```
MaxValue= broken_df[' MSSubClass'].max() where MSSubClass is
the column header
```

There are many other methods such as sort, groupby, and orderby in Pandas that are useful when playing with structured data. Also, Pandas has a ready-made adapter for popular databases such as MongoDB, Google Big Query, and so on.

One complex example with Pandas is shown next. In the X data frame for each distinct column value, find the average value of the floor grouping by the root column.

```
for col in X.columns:
                          if col != 'root':
                              avgs =
df.groupby([col,'root'],as_index=False)['floor'].
aggregate(np.mean)
                              for i,row in avgs.iterrows():
                                  k = row[col]
                                  v = row['floor']
                                  r = row['root']
                                  X.loc[(X[col] == k)
                                  & (X['root'] == r),
                                  col] = v2.
```

You can do any experiment in the Pandas framework with the data given for classification and regression problems.

ETL with Python (Unstructured Data)

Dealing with unstructured data is an important task in modern data analysis. In this section, I will cover how to parse emails, and I'll introduce an advanced research topic called *topical crawling*.

Email Parsing

See Chapter 1 for a complete example of web crawling using Python.

Like Beautiful Soup, Python has a library for email parsing. The following is the example code to parse email data stored on a mail server. The inputs in the configuration are the username and number of mails to parse for the user.

In this code, you have to mention the email user, email folder, and index of the mail-in config; code will write from the address to handle the subject and the date of the email in the CSV file.

```python
from email.parser import Parser
import os
import sys
conf = open(sys.argv[1])
config={}
users={}

# parsing the config file

for line in conf:
        if ("," in line):
                fields = line.split(",")
                key = fields[0].strip().split("=")[1].strip()
                val = fields[1].strip().split("=")[1].strip()
                users[key] = val
        else:
                if ("=" in line):
                        words = line.strip().split('=')
                        config[words[0].strip()] = words[1].strip()
conf.close()

# extracting information from user email

for usr in users.keys():
        path = config["path"]+"/"+usr+"/"+config["folder"]
        files = os.listdir(path)
        for f in sorted(files):
                if(int(f) > int(users[usr])):
                        users[usr] = f
                        path1 = path + "/" + f
```

```
                    data = ""
                    with open (path1) as myfile:
                        data=myfile.read()
                    if data != "" :
                        parser = Parser()
                    email = parser.parsestr(data)
                    out = ""
                    out = out + str(email.get('From')) + "," +
                    str(email.get('To')) + "," + str(email.get
                    ('Subject')) + "," + str(email.get
                    ('Date')).replace(","," ")
                    if email.is_multipart():
                        for part in email.get_payload():
                            out = out + "," + str(part.get_
                            payload()).replace("\n"," ")
                            .replace(","," ")
                    else:
                        out = out + "," + str(email.get_
                        payload()).replace("\n"," ").
                        replace(","," ")
                    print out,"\n"

#updating the output file

conf = open(sys.argv[1],'w')
conf.write("path=" + config["path"] + "\n")
conf.write("folder=" + config["folder"] + "\n")
for usr in users.keys():
        conf.write("name="+ usr +",value=" + users[usr] + "\n")
conf.close()
```

Sample config file for above code.

```
path=/cygdrive/c/share/enron_mail_20110402/enron_
mail_20110402/maildir
folder=Inbox
name=storey-g,value=142
name=ybarbo-p,value=775
name=tycholiz-b,value=602
```

Topical Crawling

Topical crawlers are intelligent crawlers that retrieve information from anywhere on the Web. They start with a URL and then find links present in the pages under it; then they look at new URLs, bypassing the scalability limitations of universal search engines. This is done by distributing the crawling process across users, queries, and even client computers. Crawlers can use the context available to infinitely loop through the links with a goal of systematically locating a highly relevant, focused page.

Web searching is a complicated task. A large chunk of machine learning work is being applied to find the similarity between pages, such as the maximum number of URLs fetched or visited.

Crawling Algorithms

Figure 2-2 describes how the topical crawling algorithm works with its major components.

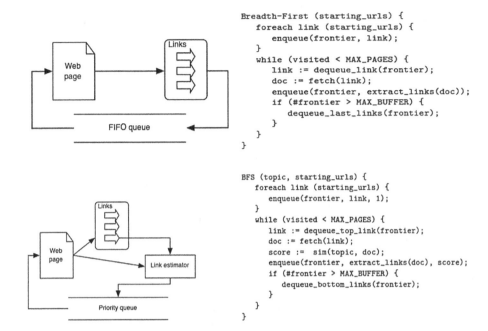

```
Breadth-First (starting_urls) {
    foreach link (starting_urls) {
        enqueue(frontier, link);
    }
    while (visited < MAX_PAGES) {
        link := dequeue_link(frontier);
        doc := fetch(link);
        enqueue(frontier, extract_links(doc));
        if (#frontier > MAX_BUFFER) {
            dequeue_last_links(frontier);
        }
    }
}
```

```
BFS (topic, starting_urls) {
    foreach link (starting_urls) {
        enqueue(frontier, link, 1);
    }
    while (visited < MAX_PAGES) {
        link := dequeue_top_link(frontier);
        doc := fetch(link);
        score := sim(topic, doc);
        enqueue(frontier, extract_links(doc), score);
        if (#frontier > MAX_BUFFER) {
            dequeue_bottom_links(frontier);
        }
    }
}
```

Figure 2-2. *Topical crawling described*

The starting URL of a topical crawler is known as the *seed URL*. There is another set of URLs known as the *target URLs*, which are examples of desired output.

Another intriguing application of crawling is for a startup that wants to uncover crucial keywords for every IP address. In the HTTP packet header, they acquire the user's browsing history from the Internet service provider. After crawling the URL visited by that IP, they classify the words in the text using name-entity recognition (Stanford NLP), which is easily implementable by the RNN explained in Chapter 5. All name entities and their types, such as names of people, locations, and organizations, are recommended for the user.

```
import requests
from bs4 import BeautifulSoup
import nltk
```

```python
from nltk.tokenize import word_tokenize
from nltk.corpus import stopwords
from nltk.tag import StanfordNERTagger
import re
import json
import os
import socket
import struct

def ip2int(addr):
    return struct.unpack("!I", socket.inet_aton(addr))[0]
def int2ip(addr):
    return socket.inet_ntoa(struct.pack("!I", addr))
java_path = '/usr/bin/java'
os.environ['JAVAHOME'] = java_path
os.environ['STANFORD_MODELS'] = '/home/ec2-user/
stanford-ner.jar'

nltk.internals.config_java(java_path)

f = open("/home/ec2-user/data.csv")

res = []

stop_words = set(stopwords.words('english'))

for line in f:
    fields = line.strip().split(",")
    url = fields[1]
    ip = fields[-1]
    print(ip)
    print(url)
    tags_del = None
    if True:
```

```
try:
    ip = ip2int(ip)
except:
    continue
print(ip)
tagged = None
try:
    code = requests.get(url)
    plain = code.text
    s = BeautifulSoup(plain)
    tags_del = s.get_text()
    if tags_del is None:
        continue
    no_html = re.sub('<[^>]*>', '', tags_del)
    st = StanfordNERTagger('/home/ec2-user/english.
    all.3class.distsim.crf.ser.gz',
                    '/home/ec2-user/stanford-ner.jar')
    tokenized = word_tokenize(no_html)
    tagged = st.tag(tokenized)
except:
    pass
if tagged is None:
    continue
for t in tagged:
    t = list(t)
    t[0] = t[0].replace(' ', '')
    t[-1] = t[-1].replace(' ', '')
    print(t)
    if t[0] in stop_words:
        continue
    unit = {}
```

```
                unit["ip"] = ip
                unit["word"] = t[0]
                unit["name_entity"] = t[-1]
                res.append(unit)

res_final = {}
res_final["result"] = res

    #except:
     #    pass
    #except:
     #    pass

with open('result.json', 'w') as fp:
    json.dump(res, fp)
```

Summary

In this chapter, we discussed different kind of databases and their use cases, and we discussed collecting text data from the Web and extracting information from different types of unstructured data like email and web pages.

CHAPTER 3

Feature Engineering and Supervised Learning

Supervised learning, also known as supervised machine learning, is a subcategory of machine learning and artificial intelligence that will help us solve a variety of real-world problems with our model. In this chapter, We will introduce the following three essential components of supervised machine learning (clustering is covered in the following chapter on unsupervised leaning), along with feature engineering, which is essential to get accurate information for the model:

- *Dimensionality reduction* tells how to choose the most important features from a set of features.

- *Classification* tells how to categorize data to a set of target categories with the help of a given training/example data set.

- *Regression* tells how to realize a variable as a linear or nonlinear polynomial of a set of independent variables.

These contents are building blocks to build any predictive model.

© Sayan Mukhopadhyay, Pratip Samanta 2023
S. Mukhopadhyay and P. Samanta, *Advanced Data Analytics Using Python*,
https://doi.org/10.1007/978 1 4842 8005 8_3

Dimensionality Reduction with Python

Dimensionality reduction is an important aspect of data analysis. It is required for both numerical and categorical data. Survey or factor analysis is one of the most popular applications of dimensionality reduction. As an example, suppose that an organization wants to find out which factors are most important in influencing or bringing about changes in its operations. It takes opinions from different employees in the organization and, based on this survey data, does a factor analysis to derive a smaller set of factors in conclusion.

In investment banking, different indices are calculated as a weighted average of instruments. Thus, when an index goes high, it is expected that instruments in the index with a positive weight will also go high and those with a negative weight will go low. The trader trades accordingly. Generally, indices consist of a large number of instruments (more than ten). In high-frequency algorithmic trading, it is tough to send so many orders in a fraction of a second. Using principal component analysis, traders realize the index as a smaller set of instruments to commence with the trading. Singular value decomposition is a popular algorithm that is used both in principal component analysis and in factor analysis. In this chapter, I will discuss it in detail. Dimensionality reduction is also required for categorical data. Suppose a retailer wants to know whether a city is an important contributor to sales volume; this can be measured by using mutual information, which will also be covered in this chapter. Before that, I will cover the Pearson correlation, which is simple to use. That's why it is a popular method of feature engineering, more precisely, as feature selection.

Correlation Analysis

There are different measures of correlation. I will limit this discussion to the Pearson correlation only. For two variables, x and y, the Pearson correlation is as follows:

$$r = \frac{\sum_i (x_i - \bar{x})(y_i - \bar{y})}{\sqrt{\sum_i (x_i - \bar{x})^2} \sqrt{\sum_i (y_i - \bar{y})^2}}$$

The value of r will vary from -1 to +1. The formula clearly shows that when x is greater than its average, then y is also greater, and therefore the r value is bigger. In other words, if x increases, then y increases, and then r is greater. So, if r is nearer to 1, it means that x and y are positively correlated. Similarly, if r is nearer to -1, it means that x and y are negatively correlated. Likewise, if r is nearer to 0, it means that x and y are not correlated. A simplified formula to calculate r is shown here:

$$r = \frac{n(\sum xy) - (\sum x)(\sum y)}{\sqrt{\left[n \sum x^2 - (\sum x)^2 \right]\left[n \sum y^2 - (\sum y)^2 \right]}}$$

You can easily use correlation for feature selection or discard similar features. Let's say Y is a variable that is a weighted sum of n variables: X1, X2, ... Xn. You want to reduce this set of X to a smaller set. To do so, you need to calculate the correlation coefficient for each X pair. Now, if Xi and Xj are highly correlated, then you will investigate the correlation of Y with

Xi and Xj. If the correlation of Xi is greater than Xj, then you remove Xj from the set, and vice versa. The following is the code for finding the correlation between two variables:

```
import numpy as np
import matplotlib.pyplot as plt
x = np.array([-4, -1, 0, 1, 3])
y = np.array([4, 1, -2, 1, 2])
corrcoef = np.corrcoef(x, y)

print(corrcoef)
```

In the output, the 1s in diagonal are representing the correlation with itself, e.g., *x* with *x* and *y* with *y*.

Now, the question is, what should be the threshold value of the previous correlation that, say, *X* and *Y* are correlated? A common practice is to assume that if $r > 0.5$, it means the variables are correlated, and if $r < 0.5$, then it means no correlation. One big limitation of this approach is that it does not consider the length of the data. For example, a 0.5 correlation in a set of 20 data points should not have the same weight as a 0.5 correlation in a set of 10,000 data points. To overcome this problem, a probable error concept has been introduced, as shown here:

$$\mathrm{PEr} = .674 \times \frac{1 - r^2}{\sqrt{n}}$$

r is the correlation coefficient, and *n* is the sample size.

Here, $r > 6\mathrm{PEr}$ means that *X* and *Y* are highly correlated, and if $r < \mathrm{Per}$, this means that *X* and *Y* are independent. Using this approach, you can see that even $r = 0.1$ means a high correlation when the data size is huge.

One interesting application of correlation is in product recommendations on an e-commerce site. Recommendations can identify similar users if you calculate the correlation of their common ratings for the same products. Similarly, you can find similar products by calculating the correlation of their common ratings from the same user. This approach is known as *collaborative filtering*.

Principal Component Analysis

Theoretically correlation works well for variables with Gaussian distribution, in other words, independent variables. For other scenarios, you have to use principal component analysis. Suppose you want to reduce the dimension of N variables: X1, X2, ... Xn. Let's form a matrix of $N{\times}N$ dimension where the i-th column represents the observation Xi, assuming all variables have N number of observations. Now if k variables are redundant, for simplicity you assume k columns are the same or linear combination of each other. Then the rank of the matrix will be $N\text{-}k$. So, the rank of this matrix is a measure of the number of independent variables, and the eigenvalue indicates the strength of that variable. This concept is used in principal component analysis and factor analysis. To make the matrix square, a covariance matrix is chosen. Singular value decomposition is used to find the eigenvalue.

Let Y be the input matrix of size $p{\times}q$, where p is the number of data rows and q is the number of parameters.

Then the $q{\times}q$ covariance matrix Co is given by the following:

$$Co = YTY / (q-1)$$

It is a symmetric matrix, so it can be diagonalized as follows:

$$Co = UDU^T$$

Each column of U is an eigenvector, and D is a diagonal matrix with eigenvalues λi in decreasing order on the diagonal. The eigenvectors are referred to as *principal axes* or *principal directions* of the data. Projections of the data on the principal axes called principal components are also known as *PC scores*; these can be seen as new, transformed variables. The j-th principal component is given by the j-th column of YU. The coordinates of the i-th data point in the new PC space are given by the i-th row of YU.

The following code is an example of PCA using Python. The input data file can be downloaded from the following link:

```
https://canvas.uw.edu/courses/1546077/files/92110986
```

```python
import pandas as pd
from sklearn.preprocessing import StandardScaler, normalize
from sklearn.decomposition import PCA

# Load the CSV file
df = pd.read_csv("book\\ch4\\CC_GENERAL.csv")

# preprocessing
df = df.drop('CUST_ID', axis = 1)
df.fillna(method ='ffill', inplace = True)

# Scale and normalize
ss = StandardScaler()
scaled_df = ss.fit_transform(df)
normalized_df= normalize(scaled_df)

# Reduce the dimensionality to 3
pca = PCA(n_components=3)
df_pca = pca.fit_transform(normalized_df)
df_pca = pd.DataFrame(df_pca)
df_pca.columns = ['P1', 'P2', 'P3']
```

If an integer is supplied to the PCA constructor, the number of principle components is examined, and if a decimal is passed, the proportion of variation to keep within components is considered. More on this can be found in Chapter 5. The physical significance of the principal component or eigenvector is demonstrated in the following example.

Example: Stretching of an elastic membrane

In the x_1x_2 plane with boundary circle $x_1^2 + x_2^2 = 1$, an elastic membrane is stretched such that the point $P(x_1, x_2)$ passes to the point $Q(y_1, y_2)$ shown by

$$[y] = \begin{bmatrix} y_1 \\ y_2 \end{bmatrix} = \begin{bmatrix} 5 & 3 \\ 3 & 5 \end{bmatrix} \begin{bmatrix} x_1 \\ x_2 \end{bmatrix} = [A][x]$$

The principal directions, which is the direction in which points P are extended radially only to points Q, are to be found. This means that some specific vectors [x] and [y] are needed so that

$$[y] = [A][x] = \lambda[x]$$

The following is a characteristic equation that gives the eigenvalues:

$$\left\| [A] - \lambda[I] \right\| = 0$$

$$\begin{vmatrix} 5 - \lambda & 3 \\ 3 & 5 - \lambda \end{vmatrix} = 0$$

$$(5 - \lambda)^2 - 9 = 0$$

$$(\lambda - 8)(\lambda - 2) = 0$$

Therefore, $\lambda = 8$ or 2. When $\lambda_1 = 8$, $[[A] - 8[I]][x]_1 = 0$.
The first eigenvector is

$$[x]_1 = \begin{bmatrix} x_1 \\ x_2 \end{bmatrix}_1 = \begin{bmatrix} 1 \\ 1 \end{bmatrix}$$

which means that $x_1 = x_2$. When $\lambda_2 = 2$, $[[A] - 2[I]][x]_2 = 0$.
The second eigenvector is

$$[x]_2 = \begin{bmatrix} x_1 \\ x_2 \end{bmatrix}_2 = \begin{bmatrix} 1 \\ -1 \end{bmatrix}$$

This means that $x_1 = -x_2$. The membrane is stretched by a factor of 8 and 2 in the principal directions, according to the eigenvalues. The boundary of the circular membrane becomes elliptical after stretching, with major and minor axes in eigenvector directions, as shown in Figure 3-1.

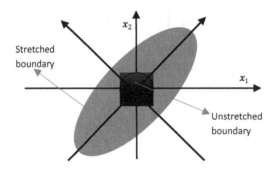

Figure 3-1. *Principle components of the stretched membrane*

Mutual Information

Mutual information (MI) of two random variables is a measure of the mutual dependence between the two variables. It is also used as a similarity measure of the distribution of two variables. A higher value of mutual information indicates the distribution of two variables is similar.

$$I(X;Y) = \sum_{x,y} p(x,y) \log \frac{p(x,y)}{p(x)p(y)}$$

Suppose a retailer wants to investigate whether a particular city is a deciding factor for its sales volume. Then the retailer can see the distribution of sales volume across the different cities. If the distribution is the same for all cities, then a particular city is not an important factor as far as sales volume is concerned. To calculate the difference between the two probability distributions, mutual information is applied here.

Here is the sample Python code to calculate mutual information:

```
from sklearn import metrics
y_true = [0, 1, 0, 1, 2, 1]
y_pred = [0, 0, 1, 1, 1, 2]
print(metrics.mutual_info_score(y_true, y_pred))
```

Let the reader have a complete code with feature engineering. It employs a deep learning model. This section covers all aspects of feature engineering.

Please find the CSV at the following link:

https://github.com/Apress/advanced-data-analytics-python-2e

```
import pandas as pd
import numpy as np
from scipy.stats.stats import pearsonr
from keras.models import Sequential
from keras.layers import Dense

#Reading input data
df = pd.read_csv('book\\ch3\\fetaure_engineering_data.csv')
df_back = pd.read_csv('book\\ch3\\fetaure_engineering_
data.csv')
print(df.shape)
```

The following code snippet normalizes the numerical columns:

```
#normalize numeric columns
mean_map = {}
std_map = {}
for col in df.columns:
    if np.issubdtype(df[col].dtype, np.number):

        avg = df[col].mean()
        st_d = df[col].std()
        df[col] = (df[col] - avg)/st_d
        mean_map[str(col)] = avg
        std_map[str(col)] = st_d
```

This part is used to transform the categorical values into numerical values:

```
#convert the categorical columns to numeric and store it

df_score = pd.DataFrame()
started = False
for col in df.columns:
    if not np.issubdtype(df[col].dtype, np.number):
        avgs = df.groupby(col, as_index=False)['SalePrice'].
        aggregate(np.mean)
        for index,row in avgs.iterrows():
            k = row[col]
            v = row['SalePrice']
            df.loc[df[col] == k, col] = v
            df_temp = pd.DataFrame()
            df_temp['feature_value'] = df_back[col]
            df_temp['score'] = df[col]
            df_temp['feature_name'] = str(col)
            df_temp = df_temp.drop_duplicates()
```

```
    if started:
        df_score = df_score.append(df_temp)
    else:
        df_score = df_temp
        started = True
df = df.fillna(df.mean())
df_score = df_score.fillna(0)
```

Here we are dropping corelated columns that are generally redundant and don't add much value in model building.

```
#drop the column which are correlated to each other

while(1):
    flag = True
    for I in range(df.shape[1]-2):
        for j in range(i+1,df.shape[1]-1):
            corr = pearsonr(df.iloc[:,i], df.iloc[:,j])
            pEr= .674 * (1- corr[0]*corr[0])/
            (len(df.iloc[:,i])**(1/2.0))
            if corr[0] > 6*pEr:
                corr1 = pearsonr(df.iloc[:,i], df''SalePric''])
                corr2 = pearsonr(df.iloc[:,j], df''SalePric''])
                stay = i

                if corr1[0]  < corr2[0]:
                    stay = j
                df.drop(df.columns[stay],inplace=True, axis=1)
                flag = False
                break
    if flag:
        break
print(df.shape)
```

Finally, we can see there are 25 columns in the data frame, whereas in the beginning we had 81 columns. With the previous code, we transformed the data and removed unnecessary columns.

Now we'll discuss the classification problems of supervised learning.

Classifications with Python

Classification is a well-accepted example of machine learning. It has a set of a target classes and training data. Each training data is labeled by a particular target class. The classification model is trained by training data and predicts the target class of test data. One common application of classification is in fraud identification in the credit card or loan approval process. It classifies the applicant as fraud or nonfraud based on data. Many uneducated shopkeepers classify their goods without doing any calculations. They did not buy a product that did not sell well the last week. It is a Bayesian classifier of some sort. The advertising industry blocks impressions with the previous week's average price below a certain threshold. Classification is also widely used in image recognition. From a set of images, if you recognize the image of a computer, it is classifying the image of a computer and not of a computer class.

Sentiment analysis is a popular application of text classification. Suppose an airline company wants to analyze its customer textual feedback. Then each feedback is classified according to sentiment (positive/negative/neutral) and also according to context (about staff/timing/food/price). Once this is done, the airline can easily find out what the strength of that airline's staff is or its level of punctuality or cost effectiveness or even its weakness. Broadly, there are three approaches in classification.

- *Rule-based approach*: I will discuss the decision tree and random forest algorithm.

- *Probabilistic approach*: I will discuss the naïve Bayes algorithm.

- *Distance-based approach*: I will discuss the support vector machine.

Semi-Supervised Learning

Classification and regression are types of supervised learning. In this type of learning, you have a set of training data where you train your model. Then the model is used to predict test data. For example, suppose you want to classify text according to sentiment. There are three target classes: positive, negative, and neutral. To train your model, you have to choose some sample text and label it as positive, negative, and neutral. You use this training data to train the model. Once your model is trained, you can apply your model to test data. For example, you may use the naïve Bayes classifier for text classification and try to predict the sentiment of the sentence "Food is good." In the training phase, the program will calculate the probability of a sentence being positive or negative or neutral when the words *Food*, *is*, and *good* are presented separately and stored in the model, and in the test phase it will calculate the joint probability when *Food*, *is*, and *good* all come together.

Conversely, clustering is an example of unsupervised learning where there is no training data or target class available. The program learns from data in one shot. There is an instance of semi-supervised learning also. Suppose you are classifying the text as positive and negative sentiments but your training data has only positives. The training data that is not positive is unlabeled. In this case, as the first step, you train the model assuming all unlabeled data is negative and apply the trained model on the training data. In the output, the data coming in as negative should be labeled as negative. Finally, train your model with the newly labeled data. The nearest neighbor classifier is also considered as semi-supervised learning. It has training data, but it does not have the training phase of the model.

Decision Tree

A *decision tree* is a tree of rules. Each level of the tree represents a parameter, each node of the level validates a constraint for that level parameter, and each branch indicates a possible value of a parent node parameter. Figure 3-2 shows an example of a decision tree.

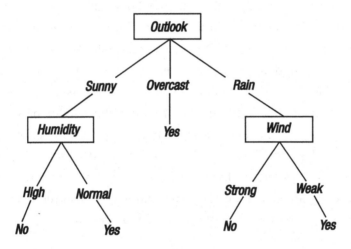

Figure 3-2. *Example of decision tree for good weather*

Which Attribute Comes First?

One important aspect of the decision tree is to decide the order of features. The entropy-based information gain measure decides it.

Entropy is a measure of randomness of a system.

$$Entropy(S) \equiv \sum_{i=1}^{c} -p_i \log_2 p_i$$

For example, for any obvious event like the sun rises in the east, entropy is zero, $P=1$, and $\log(p)=0$. More entropy means more uncertainty or randomness in the system.

Information gain, which is the expected reduction in entropy caused by partitioning the examples according to this attribute, is the measure used in this case.

Specifically, the information gain, *Gain(S,A)*, of an attribute *A* relative to a collection of examples *S* is defined as follows:

$$Gain(S,A) \equiv Entropy(S) - \sum_{v \in Values(A)} \frac{|S_v|}{|S|} Entropy(S_v)$$

So, an attribute with a higher information gain will come first in the decision tree.

```
from sklearn.datasets import load_iris
from sklearn.tree import DecisionTreeClassifier
iris_data = load_iris()
data, labels = iris_data.data, iris_data.target
model = DecisionTreeClassifier()
model = model.fit(data, labels)
print(model.predict(data[[10]]))
```

Random Forest Classifier

A *random forest classifier* is an extension of a decision tree in which the algorithm creates *N* number of decision trees where each tree has *M* number of features selected randomly. Now a test data will be classified by all decision trees and be categorized in a target class that is the output of the majority of the decision trees.

```
from sklearn.datasets import load_iris
from sklearn.ensemble import RandomForestClassifier
iris_data = load_iris()
data, labels = iris_data.data, iris_data.target
model = RandomForestClassifier()
model = model.fit(data, labels)
print(model.predict(data[[10]]))
```

Naïve Bayes Classifier

X = (x1, x2, x3, ..., xn) is a vector of n dimension. The Bayesian classifier assigns each X to one of the target classes of set {C1, C2, ..., Cm,}. This assignment is done on the basis of probability that X belongs to target class Ci. That is to say, X is assigned to class Ci if and only if P(Ci |X) > P(Cj |X) for all j such that $1 \leq j \leq m$.

$$P(C_i|X) = \frac{P(X|C_i)P(C_i)}{P(X)}$$

In general, it can be costly computationally to compute P(X|Ci). If each component x_k of X can have one of *r* values, there are r^n combinations to consider for each of the *m* classes. To simplify the calculation, the assumption of conditional class independence is made, which means that for each class, the attributes are assumed to be independent. The classifier developing from this assumption is known as the naïve Bayes classifier. The assumption allows you to write the following:

$$P\left(X|C_i\right)=\prod_{k=1}^{n}P\left(x_k|C_i\right)$$

The following code is an example of the naïve Bayes classification of numerical data:

```
from sklearn.naïve_bayes import GaussianNB
from sklearn.datasets import load_iris
iris_data = load_iris()
data, labels = iris_data.data, iris_data.target
model = GaussianNB()
model = model.fit(data, labels)
print(model.predict(data[[50]]))
```

Note   You'll see another example of the naïve Bayes classifier in the "Sentiment Analysis" section.

Support Vector Machine

If you look at Figure 3-3, you can easily understand that the circle and square points are linearly separable in two dimensions (x1, x2). But they are not linearly separable in either dimension x1 or x2. The support vector machine algorithm works on this theory. It increases the dimension of the data until points are linearly separable. Once that is done, you have to find two parallel hyperplanes that separate the data. This planes are known as the *margin*. The algorithm chose the margins in such a way that the distance between them is the maximum. That's why it is the maximum margin. The plane, which is at the middle of these two margins or at equal distance between them, is known as an *optimal hyperplane* that is used to classify the test data (see Figure 3-2). The separator can be nonlinear also.

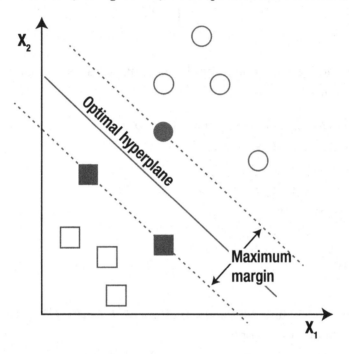

Figure 3-3. *Support vector machine*

The following code is an example of doing support vector machine classification using Python:

```
from sklearn.svm import SVC
from sklearn.datasets import load_iris
iris_data = load_iris()
data, labels = iris_data.data, iris_data.target
model = SVC()
model = model.fit(data, labels)
print(model.predict(data[[50]]))
```

Nearest Neighbor Classifier

The nearest neighbor classifier is a simple distance-based classifier. It calculates the distance of test data from the training data and groups the distances according to the target class. The target class, which has a minimum average distance from the test instance, is selected as the class of the test data. A Python example is shown here:

```
from sklearn.neighbors import KNeighborsClassifier
from sklearn.datasets import load_iris
iris_data = load_iris()
data, labels = iris_data.data, iris_data.target
model = KNeighborsClassifier()
model = model.fit(data, labels)
print(model.predict(data[[50]]))
```

Sentiment Analysis

Sentiment analysis is an interesting application of text classification. For example, say one airline client wants to analyze its customer feedback. It classifies the feedback according to sentiment (positive/negative) and also by aspect (food/staff/punctuality). After that, it can easily understand its strengths in business (the aspect that has the maximum positive feedback) and its weaknesses (the aspect that has the maximum negative feedback). The airline can also compare this result with its competitor. One interesting advantage of doing a comparison with the competitor is that it nullifies the impact of the accuracy of the model because the same accuracy is applied to all competitors. This is simple to implement in Python using the textblob library, as shown here:

```
from textblob.classifiers import NaïveBayesClassifier
train = [('I love this sandwich.', 'pos'),  ('this is an
amazing place!', 'pos'),('I feel very good about these beers.',
'pos'),('this is my best work.', 'pos'),("what an awesome view",
'pos'),('I do not like this restaurant', 'neg'),('I am tired of
this stuff.', 'neg'),("I can't deal with this", 'neg'),('he is
my sworn enemy!', 'neg'),('my boss is horrible.', 'neg')]
cl = NaïveBayesClassifier(train)
print (cl.classify("This is an amazing library!"))
```

   output : pos

```
from textblob.classifiers import NaïveBayesClassifier
train = [('Air India did a poor job of queue management both
times.', 'staff service'),  ("The 'cleaning' by flight attendants
involved regularly spraying air freshener in the lavatories.",
'staff'),('The food tasted decent.', 'food'),('Flew Air India
direct from New York to Delhi round trip.', 'route'),('Colombo
to Moscow via Delhi.', 'route'),('Flew Birmingham to Delhi with
Air India.', 'route'),('Without toilet, food or anything!',
'food'),('Cabin crew announcements included a sincere apology
for the delay.', 'cabin flown')]
cl = NaïveBayesClassifier(train)
tests = ['Food is good.']
for c in tests:
      print(cl.classify(c))
```

The textblob library also supports a random forest classifier, which works best on text written in proper English such as a formal letter might be written. For text that is not usually written with proper grammar, such as customer feedback, naïve Bayes works better. Naïve Bayes has another advantage in real-time analytics. You can update the model without losing the previous training.

Image Recognition

Image recognition is a common example of image classification. It is easy to use in Python by applying the opencv library.

Please download the model file (haarcascade_frontalface_default.xml) from OpenCV GitHub (https://github.com/opencv/opencv/tree/master/data/haarcascades) as well as the sample Lenna image from https://en.wikipedia.org/wiki/File:Lenna_(test_image).png.

Here is the sample code:

```
import cv2
# Load the cascade
face_model = cv2.CascadeClassifier('haarcascade_frontalface_
default.xml')
# Reading the image
img = cv2.imread('lena.jpg')
# Convert into grayscale
gray_img = cv2.cvtColor(img, cv2.COLOR_BGR2GRAY)
# Detect faces
faces_detected = face_model.detectMultiScale(gray_img, 1.1, 4)
# Drawing rectangle around the faces
for (x, y, w, h) in faces_detected:
    cv2.rectangle(img, (x, y), (x+w, y+h), (255, 0, 0), 2)
# Show the output
cv2.imshow('img', img)
cv2.waitKey()

Install opencv with command:
sudo pip3 install opencv-contrib-python
```

Now we will discuss another type of supervised learning regression problem.

Regression with Python

Regression realizes a variable as a linear or nonlinear polynomial of a set of independent variables.

Here is an interesting use case: what is the sales price of a product that maximizes its profit? This is a million-dollar question for any merchant. The question is not straightforward. Maximizing the sales price may not result in maximizing the profit because increasing the sales price sometimes decreases the sales volume, which decreases the total profit. So, there will be an optimized value of sales price for which the profit will be at the maximum.

There are N records of the transaction with M number of features called F1, F2, ... Fm (sales price, buy price, cash back, SKU, order date, and so on). You have to find a subset of $K(K<M)$ features that have an impact on the profit of the merchant and suggest an optimal value of V1, V2, ... Vk for these K features that maximize the revenue.

You can calculate the profit of merchant using the following formula:

$$\text{Profit} = (\text{SP-TCB-BP})^* \, \text{SV} \qquad (1)$$

For this formula, here are the variables:

- SP = Sales price

- TCB = Total cash back

- BP = Buy price

- SV = Sales volume

Now using regression, you can realize SV as follows:

$$\text{SV} = a + b^*\text{SP} + c^*\text{TCB} + d^*\text{BP}$$

Now you can express profit as a function of SP, TCB, and BP and use mathematical optimization. With constraining in all parameter values, you can get optimal values of the parameters that maximize the profit.

This is an interesting use case of regression. There are many scenarios where one variable has to be realized as the linear or polynomial function of other variables.

Least Square Estimation

Least square estimation is the simplest and oldest method for doing regression. It is also known as the *curve fitting method*. Ordinary least squares (OLS) regression is the most common technique and was invented in the 18th century by Carl Friedrich Gauss and Adrien-Marie Legendre. The following is a derivation of coefficient calculation in ordinary least square estimation:

$$Y = X\beta + \in$$
$$X'Y = X'X\beta + X'\in$$
$$X'Y = X'X\beta + 0$$
$$(X'X)^{-1} X'Y = \beta + 0$$
$$\beta = (X'X)^{-1} X'Y$$

The following code is a simple example of OLS regression:

Please find the CSV file at the following link:

https://github.com/Apress/advanced-data-analytics-python-2e

```
import pandas as pd
import statsmodels.api as sm
df = pd.read_csv('book\\ch3\\fetaure_engineering_data.csv')
# df = pandas.read_csv("restaurants.csv")
X = df['LotArea']
Y = df['SalePrice']
X = sm.add_constant(X)
model = sm.OLS(Y, X).fit()
summary = model.summary()
print(summary)
```

Logistic Regression

Logistic regression is an interesting application of regression that calculates the probability of an event. It introduces an intermediate variable that is a linear regression of linear variable. Then it passes the intermediate variable through the logistic function, which maps the intermediate variable from zero to one. The variable is treated as a probability of the event.

The following code is an example of doing logistic regression on numerical data:

```
from sklearn.preprocessing import StandardScaler
from sklearn.linear_model import LogisticRegression
from sklearn.datasets import load_iris
data, labels = load_iris(return_X_y=True)
scaler = StandardScaler()
data = scaler.fit_transform(data)
model = LogisticRegression(random_state=0)
model = model.fit(data, labels)
print(model.predict(data[[50]]))
```

Classification and Regression

Classification and regression may be applied on the same problem. For example, when a bank approves a credit card or loan, it calculates a credit score of a candidate based on multiple parameters using regression and then sets up a threshold. Candidates having a credit score greater than the threshold are classified as potential nonfraud, and the remaining are considered as potential fraud. Likewise, any binary classification problem can be solved using regression by applying a threshold on the regression result. In Chapter 4, I discussed in detail how to choose a threshold value from the distribution of any parameter. Similarly, some binary classifications can be used in place of logistic regression. For instance, say an e-commerce site wants to predict the chances of a purchase order being converted into a successful invoice. The site can easily do so using logistic regression. The naïve Bayes classifier can be directly applied on the problem because it calculates probability when it classifies the purchase order to be in the successful or unsuccessful class. The random forest algorithm can be used in the problem as well. In that case, among the N decision tree, if the M decision tree says the purchase order will be successful, then M/N will be the probability of the purchase order being successful.

Intentionally Bias the Model to Over-Fit or Under-Fit

Sometimes you need to over- or under-predict intentionally. In an auction, when you are predicting from the buy side, it will always be good if your bid is a little lower than the original. Similarly, on the sell side, it is desired that you set the price a little higher than the original. You can do this in two ways. In regression, when you are selecting the features using correlation, over-predicting intentionally drops some variable with negative correlation. Similarly, under-predicting drops some variable with positive correlation. There is another way of dealing with this. When you are predicting the value, you can predict the error in the prediction. To over-predict, when you see that the predicted error is positive, reduce the prediction value by the error amount. Similarly, to over-predict, increase the prediction value by the error amount when the error is positive.

Another problem in classification is biased training data. Suppose you have two target classes, A and B. The majority (say 90 percent) of training data is class A. So, when you train your model with this data, all your predictions will become class A. One solution is a biased sampling of training data. Intentionally remove the class A example from training. Another approach can be used for binary classification. As class B is a minority in the prediction probability of a sample, in class B it will always be less than 0.5. Then calculate the average probability of all points coming into class B. For any point, if the class B probability is greater than the average probability, then mark it as class B and otherwise class A.

Dealing with Categorical Data

For algorithm-like support, vector or regression input data must be numeric. So, if you are dealing with categorical data, you need to convert to numeric data. One strategy for conversion is to use an ordinal number as the numerical score. Other popular technique is using *one-hot encoder*. In the first method, we assign numbers for each of the category. If there are five categories in a column, there will be five numbers, e.g., 0 to 4, for these categories, whereas in one-hot encoding, a new binary column is created for each category and marked as true or false based on the category. Please run the following code to get a better understanding of this:

Please find the CSV file at the following link:

https://github.com/Apress/advanced-data-analytics-python-2e

```
import pandas as pd
from sklearn.preprocessing import OneHotEncoder
from sklearn.preprocessing import OrdinalEncoder

df = pd.read_csv('book\\ch3\\fetaure_engineering_data.csv')

oe = OrdinalEncoder()
df["Functional_code"] = oe.fit_transform(df[["Functional"]])
print(df[["Functional", "Functional_code"]]. head(20))

ohe = OneHotEncoder()
ohe_results = ohe.fit_transform(df[["Functional"]])
print(pd.DataFrame(ohe_results.toarray(), columns=ohe.
categories).head(20))
```

Summary

This chapter discusses feature engineering, which is essential to get good accuracy for the model. Then we describe supervised learning techniques like classification and regression in detail.

CHAPTER 4

Unsupervised Learning: Clustering

In Chapter 3 we discussed how training data can be used to categorize customer comments according to sentiment (positive, negative, neutral), as well as according to context. For example, in the airline industry, that context might be punctuality, food, comfort, entertainment, and so on. Using this analysis, a business owner can determine the areas that his business needs to concentrate on. For instance, if he observes that the highest percentage of negative comments has been about food, then his priority will be the quality of food being served to customers. However, there are scenarios where business owners are not sure about the context. There are also cases where training data is not available. Moreover, the frame of reference can change with time. Classification algorithms cannot be applied where target classes are unknown. A clustering algorithm is used in these kinds of situations. A conventional application of clustering is found in the wine-making industry where each cluster represents a brand of wine, and wines are clustered according to their component ratio in wine. In Chapter 3, you learned that classification can be used to recognize a type of image, but there are scenarios where one image has multiple shapes and an algorithm is needed to separate the figures. Clustering algorithms are used in this kind of use case.

© Sayan Mukhopadhyay, Pratip Samanta 2023
S. Mukhopadhyay and P. Samanta, *Advanced Data Analytics Using Python*,
https://doi.org/10.1007/978-1-4842-8004-1_4

Clustering classifies objects into groups based on similarity or distance measure. This is an example of unsupervised learning. The main difference between clustering and classification is that the latter has well-defined target classes. The characteristics of target classes are defined by the training data and the models learned from it. That is why classification is supervised in nature. In contrast, clustering tries to define meaningful classes based on data and its similarity or distance. Figure 4-1 illustrates a document clustering process.

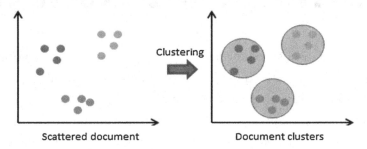

Figure 4-1. *Document clustering*

K-Means Clustering

Let's suppose that a retail distributer has an online system where local agents enter trading information manually. One of the fields they have to fill in is City. But because this data entry process is manual, people normally tend to make lots of spelling errors. For example, instead of Delhi, people enter *Dehi, Dehli, Delly*, and so on. You can try to solve this problem using clustering because the number of clusters is already known; in other words, the retailer is aware of how many cities the agents operate in. This is an example of *K-means clustering*.

The K-means algorithm has two inputs. The first one is the data X, which is a set of N number of vectors, and the second one is K, which represents the number of clusters that need to be created. The output is a set of K centroids in each cluster as well as a label to each vector in X that indicates the points assigned to the respective cluster. All points within a cluster are nearer in distance to their centroid than any other centroid. The condition for the K clusters C_k and the K centroids μ_k can be expressed as follows:

$$\text{minimize} \sum_{k=1}^{K} \sum_{x_n \in C_k} \|X_n - \mu_k\|^2 \text{ with respect to } C_k, \mu_k.$$

However, this optimization problem cannot be solved in polynomial time. But Lloyd has proposed an iterative method as a solution. It consists of two steps that iterate until the program converges to the solution.

1. It has a set of K centroids, and each point is assigned to a unique cluster or centroid, where the distance of the concerned centroid from that particular point is the minimum.

2. It recalculates the centroid of each cluster by using the following formula:

$$C_k = \left\{ Xn : \|Xn - \mu k\| \le \text{all } \|Xn - \mu l\| \right\} \qquad (1)$$

$$\mu_k = \frac{1}{Ck} \sum_{Xn \in Ck} Xn \qquad (2)$$

The two-step procedure continues until there is no further rearrangement of cluster points. The convergence of the algorithm is guaranteed, but it may converge to a local minima.

The following is a simple implementation of Lloyd's algorithm for performing K-means clustering in Python:

```python
import random
def ED(source, target):
        if source == "":
                return len(target)
```

```
    if target == "":
        return len(source)
    if source[-1] == target[-1]:
        cost = 0
    else:
        cost = 1
    res = min([ED(source[:-1], target)+1,ED(source,
    target[:-1])+1,ED(source[:-1], target[:-1]) + cost])
    Return(res)

def find_centre(x, X, mu):
    min = 100
    cent = 0
    for c in mu:
        dist = ED(x, X[c])
        if dist< min:
            min = dist
            cent = c
    return(cent)

def cluster_arrange(X, cent):
    clusters  = {}
    for x in X:
    bestcent = find_centre(x, X, cent)
    try:
        clusters[bestcent].append(x)
    exceptKeyError:
        clusters[bestcent] = [x]
    return( clusters)
```

```
def rearrange_centers(cent, clusters):
    newcent = []
    keys = sorted(clusters.keys())
    for k in keys:
        newcent.append(k)
    return(newcent)
def has_converged(cent, oldcent):
    return (sorted(cent) == sorted(oldcent))
def locate_centers(X, K):
    oldcent = random.sample(range(0,5), K)
    cent = random.sample(range(0,5), K)
    while not has_converged(cent, oldcent):
        oldcent = cent
            # Assign all points in X to clusters
        clusters = cluster_arrange(X, cent)
            # Reevaluate centers
        cent = rearrange_centers(oldcent, clusters)
return(cent, clusters)
X = ['Delhi','Dehli', 'Delli','Kolkata','Kalkata','Kalkota']
print(locate_centers(X,2))
```

However, K-means clustering has a limitation. For example, suppose all of your data points are located in eastern India. For K=4 clustering, the initial step is that you randomly choose a center in Delhi, Mumbai, Chennai, and Kolkata. All of your points lie in eastern India, so all the points are nearest to Kolkata and are always assigned to Kolkata. Therefore, the program will converge in one step. To avoid this problem, the algorithm is run multiple times and averaged. Programmers can use various tricks to initialize the centroids in the first step.

Choosing K: The Elbow Method

There are certain cases where you have to determine the *K* in K-means clustering. For this purpose, you have to use the elbow method, which uses a percentage of variance as a variable dependent on the number of clusters. Initially, several clusters are chosen. Then another cluster is added, which doesn't make the modeling of data much better. The number of clusters is chosen at this point, which is the *elbow criterion*. This "elbow" cannot always be unambiguously identified. The percentage of variance is realized as the ratio of the between-group variance of individual clusters to the total variance. Assume that in the previous example, the retailer has four cities: Delhi, Kolkata, Mumbai, and Chennai. The programmer does not know that so does clustering with K=2 to K=9 and plots the percentage of variance. The programmer will get an elbow curve that clearly indicates K=4 is the right number for K.

Silhouette Analysis

The silhouette analysis is another method of choosing the K value. It measures how similar an object is to its own cluster (cohesion) compared to other clusters (separation). The range of silhouette value is −1 to +1; the higher the value, the more the object is matched to its own cluster. It is desirable to have a high value for most data points. If many points have a low or negative value, then the clustering configuration is not right.

The silhouette can be calculated with any distance metric, such as the Euclidean distance, which is discussed later.

The Silhouette score s(i) for each data point i is defined as follows:

$$s(i) = \frac{b(i) - a(i)}{\max\{a(i), b(i)\}}, \text{if} |C_i| > 1$$

and

$$s(i) = 0, \text{if} |C_i| = 1$$

Source: Wikipedia

Here, $a(i)$ is the measure of similarity of the point i to its own cluster, $b(i)$ is the measure of dissimilarity of i from points in other clusters, and $s(i)$ is defined to be equal to zero if i is the only point in the cluster.

Please refer to the following code for an example of silhouette scoring:

```
from sklearn.datasets import make_blobs
from sklearn.cluster import KMeans
import matplotlib.pyplot as fig
from sklearn.metrics import silhouette_score

# Create dataset with 4 random cluster centers and 1000
datapoints
x, y = make_blobs(n_samples = 1000, centers = 4, n_features=2,
shuffle=True, random_state=31)

sil_score = []
kmax = 10

# number of clusters should be greater or equal to 2
for k in range(2, kmax+1):
  kmeans = KMeans(n_clusters = k).fit(x)
  labels = kmeans.labels_
  sil_score.append(silhouette_score(x, labels, metric =
  'euclidean'))
```

```
fig.plot([i for i in range(2,11)], sil_score)
fig.xlabel("K value")
fig.ylabel("silhouette_score")
fig.show()
```

Figure 4-2 is generated by the code. This clearly depicts that the silhouette score is maximum for the k value 4. So, we need to choose four clusters.

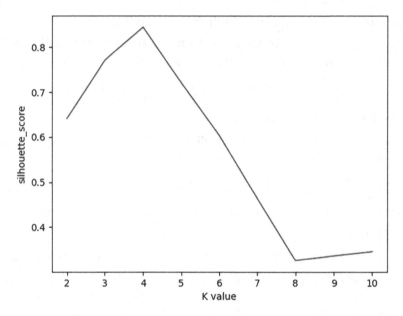

Figure 4-2. *Silhouette score versus k value*

Distance or Similarity Measure

The measure of distance or similarity is one of the key factors of clustering. In this section, we will describe the different kinds of distance and similarity measures. Before that, we'll explain what *distance* actually means here.

Properties

The distances are measures that satisfy the following properties:

- dist(x, y) = 0 if and only if x=y.

- dist(x, y) > 0 when x ≠ y.

- dist(x, y) = dist(x, y).

- dist(x,y) + dist(y,z) >= d(z,x) for all x, y, and z.

General and Euclidean Distance

The distance between the points p and q is the length of the geometrical line between them: ($\overline{pq}$). This is called *Euclidean distance*.

According to Cartesian coordinates, if $p = (p_1, p_2,..., p_n)$ and $q = (q_1, q_2,..., q_n)$ are the two points in Euclidean n-space, then the distance (d) from q to p or from p to q is derived from the Pythagorean theorem, as shown here:

$$d(p,q) = d(q,p) = \sqrt{(q_1 - p_1)^2 + (q_2 - p_2)^2 + ... + (q_n - p_n)^2}$$
$$= \sqrt{\sum_{i=1}^{n}(q_i - p_i)^2}.$$

The Euclidean vector is the position of a point in a Euclidean n-space. The magnitude of a vector is a measure. Its length is calculated by the following formula:

$$\|p\| = \sqrt{p_1^2 + p_2^2 + \cdots + p_n^2} = \sqrt{p \cdot p},$$

A vector has direction along with a distance. The distance between two points, p and q, may have a direction, and therefore, it may be represented by another vector, as shown here:

$$q - p = \left(q_1 - p_1, q_2 - p_2, \ldots, q_n - p_n \right)$$

The Euclidean distance between p and q is simply the Euclidean length of this distance (or displacement) vector.

$$\|q - p\| = \sqrt{(q - p) \cdot (q - p)}$$
$$\|q - p\| = \sqrt{\|p\|^2 + \|q\|^2 - 2p \cdot q}$$

In one dimension:

$$\sqrt{(x - y)^2} = |x - y|.$$

In two dimensions:

In the Euclidean plane, if $p = (p_1, p_2)$ and $q = (q_1, q_2)$, then the distance is given by the following:

$$d(p, q) = \surd (q_1 - p_1)\wedge 2 + (q_2 - p_2)\wedge 2$$

Alternatively, it follows from the equation that if the polar coordinates of the point p are (r_1, θ_1) and those of q are (r_2, θ_2), then the distance between the points is as follows:

$$\surd r_1 \wedge 2 + r_2 \wedge 2 - 2 r_1 r_2 \cos\left(\theta_1 - \theta_2 \right)$$

In n dimensions:

In the general case, the distance is as follows:

$$D^2(p,q) = \vdash (p_1 - q_1)^2 + (p_2 - q_2)^2 + \ldots + (p_i - q_i)^2 + \ldots + (p_n - q_n)^2.$$

In Chapter 3, you will find an example of Euclidian distance in the nearest neighbor classifier example.

Squared Euclidean Distance

The standard Euclidean distance can be squared to place progressively greater weight on objects that are farther apart. In this case, the equation becomes the following:

$$d^2(p,q) = \vdash (p_1 - q_1)^2 + (p_2 - q_2)^2 + \ldots + (p_i - q_i)^2 + \ldots + (p_n - q_n)^2.$$

Squared Euclidean distance is not a metric because it does not satisfy the triangle inequality. However, it is frequently used in optimization problems in which distances are to be compared only.

Distance Between String-Edit Distance

Edit distance is a measure of dissimilarity between two strings. It counts the minimum number of operations required to make two strings identical. Edit distance finds applications in natural language processing, where automatic spelling corrections can indicate candidate corrections for a misspelled word. Edit distance is of two types.

- Levenshtein edit distance

- Needleman edit distance

Levenshtein Distance

The Levenshtein distance between two words is the least number of insertions, deletions, or replacements that need to be made to change one word into another. In 1965, it was Vladimir Levenshtein who considered this distance.

Levenshtein distance is also known as *edit distance*, although that might denote a larger family of distance metrics as well. It is affiliated with pair-wise string alignments.

For example, the Levenshtein distance between Calcutta and Kolkata is 5, since the following five edits change one into another:

Calcutta →→ Kalcutta (substitution of *C* for *K*)

Kalcutta →→ Kolcutta (substitution of *a* for *o*)

Kolcutta →→ Kolkutta (substitution of *c* for *k*)

Kolkutta →→ Kolkatta (substitution of *u* for *a*)

Kolkatta →→ Kolkata (deletion of *t*)

When the strings are identical, the Levenshtein distance has several simple upper bounds that are the lengths of the larger strings, and the lower bounds are zero. The code example of the Levinstein distance is given in the K-mean clustering code.

Needleman–Wunsch Algorithm

The Needleman–Wunsch algorithm is used in bioinformatics to align protein or nucleotide sequences. It was one of the first applications of dynamic programming for comparing biological sequences. It works using dynamic programming. First it creates a matrix where the rows and columns are letters. Each cell of the matrix is a similarity score of the corresponding letter in that row and column. Scores are one of three types: matched, not matched, or matched with insert or deletion. Once the matrix is filled, the algorithm does a backtracing operation from the bottom-right cell to the top-left cell and finds the path where the neighbor score distance is the minimum. The sum of the score of the backtracing path is the Needleman–Wunsch distance for two strings.

Pyopa is a Python module that provides a ready-made Needleman–Wunsch distance between two strings.

```python
import pyopa
data = {'gap_open': -20.56,
        'gap_ext': -3.37,
        'pam_distance': 150.87,
        'scores': [[10.0]],
        'column_order': 'A',
        'threshold': 50.0}
env = pyopa.create_environment(**data)
s1 = pyopa.Sequence('AAA')
s2 = pyopa.Sequence('TTT')
print(pyopa.align_double(s1, s1, env))
print(env.estimate_pam(aligned_strings[0], aligned_strings[1]))
```

Although Levenshtein is simple in implementation and computationally less expensive, if you want to introduce a gap in string matching (for example, *New Delhi* and *NewDelhi*), then the Needleman–Wunsch algorithm is the better choice.

Similarity in the Context of a Document

A similarity measure between documents indicates how identical two documents are. Generally, similarity measures are bounded in the range [-1,1] or [0,1] where a similarity score of 1 indicates maximum similarity.

Types of Similarity

To measure similarity, documents are realized as a vector of terms excluding the stop words. Let's assume that A and B are vectors representing two documents. In this case, the different similarity measures are shown here:

- **Dice**

    The Dice coefficient is denoted by the following:

    $$\text{sim}(q,d_j) = D(A,B) = \frac{|A \cap B|}{\alpha|A| + (1-\alpha)|B|}$$

    Also,

    $\alpha \in [0, 1]$ and let $\alpha = \dfrac{1}{2}$

- **Overlap**

    The Overlap coefficient is computed as follows:

    $$\text{Sim}(q,d_j) = O(A,B) = \frac{|A \cap B|}{\min(|A|,|B|)}$$

    The Overlap coefficient is calculated using the max operator instead of min.

- **Jaccard**

    The Jaccard coefficient is given by the following:

    $$\text{Sim}(q,d_j) = J(A,B) = \frac{|A \cap B|}{|A \cup B|}$$

The Jaccard measure signifies the degree of relevance.

- **Cosine**

  The cosine of the angle between two vectors is given by the following:

$$Sim(q,d_j) = C(A,B) = \frac{|A \cap B|}{\sqrt{|A||B|}}$$

Distance and similarity are two opposite measures. For example, numeric data correlation is a similarity measure, and Euclidian distance is a distance measure. Generally, the value of the similarity measure is limited to between 0 and 1, but distance has no such upper boundary. Similarity can be negative, but by definition, distance cannot be negative. The clustering algorithms are almost the same as from the beginning of this field, but researchers are continuously finding new distance measures for varied applications.

In the next section, we'll use the K-means algorithm to cluster shades of colors in an image now that we've gone through the basics of the technique.

Example of K-Means in Images

To begin with, we'll be given a canvas of blobs, each of which is a different hue of red, green, or blue (Figure 4-3).

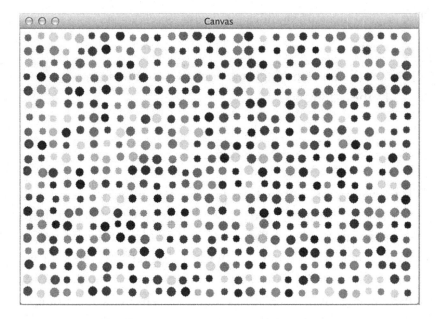

Figure 4-3. *A red, green, and blue blobbed canvas*

The aim is to divide the blobs into three categories based on their hue. Then we'll use K-means to group the red, green, and blue blobs together while separating them. Figure 4-4 shows a sample output image in which the red blobs have been isolated from all other blobs.

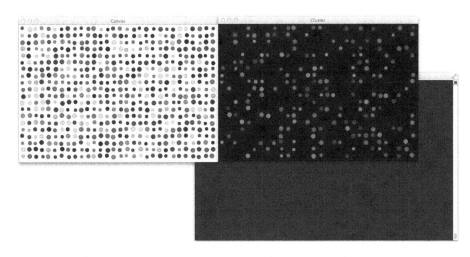

Figure 4-4. *Separating the red blobs from the green and blue blobs*

Preparing the Cluster

The first step is to create a picture that looks similar to Figure 4-2 so that we can cluster it as follows:

```
# import the necessary packages
from sklearn.cluster import KMeans
import numpy as np
import random
import cv2
import imutils

# initialize the list of hue choices
hues = [
        # shades of red, green, and blue
        (138, 8, 8), (180, 4, 4), (223, 1, 1), (255, 0, 0),
        (250, 88, 88),
        (8, 138, 8), (4, 180, 4), (1, 223, 1), (0, 255, 0),
        (46, 254, 46),
```

```
    (11, 11, 97), (8, 8, 138), (4, 4, 180), (0, 0, 255),
    (46, 46, 254)]

# initialize the canvas
canvas = np.ones((400, 600, 3), datatype="uint8") * 255

# loop over the canvas
for y in range(0, 400, 20):
    for x in range(0, 600, 20):
        # generate a random (x, y) coordinate, radius,
        and hue for
        # the circle
        (dx, dy) = np.random.randint(5, 10, size=(2,))
        r = np.random.randint(5, 8)
        hue = random.choice(hues)[::-1]

        # draw the circle on the canvas
        cv2.circle(canvas, (x + dx, y + dy), r, hue, -1)

# pad the border of the image
canvas = cv2.copyMakeBorder(canvas, 5, 5, 5, 5, cv2.BORDER_
CONSTANT,
    value=(255, 255, 255))
```

Lines 9–13 of the previous code set up five different colors of red, green, and blue. When we draw our colors on our canvas on line 16, we'll take a random sample from this list. Lines 19–28 deal with the actual procedure of drawing the random blobs. We begin by looping through the entire image in increments of 20 pixels. Then, in lines 23 and 24, we produce a random (x, y)-coordinate for the blob's center (line 23), select a random radius for the blob (line 24), and sample from our colors list on line 25. The hue list is in RGB order; we'll invert it to BGR order so that OpenCV can draw it appropriately. Lastly, we use the randomly generated values to draw a circle on our canvas. Figure 4-2 shows the output of this code.

Thresholding

We need to identify each blob and extract color characteristics to describe them now that we have our artificial image of blobs. Here's an example using thresholding, for example:

```
# convert the canvas to grayscale, threshold it, and detect
# contours in the image
gray = cv2.cvtHue(canvas, cv2.HUE_BGR2GRAY)
gray = cv2.bitwise_not(gray)
thresh = cv2.threshold(gray, 10, 255, cv2.THRESH_BINARY)[1]
cnts = cv2.findContours(gray.copy(), cv2.RETR_LIST, cv2.CHAIN_
APPROX_SIMPLE)
cnts = imutils.grab_contours(cnts)

# initialize the data matrix
data = []

# loop over the contours
for c in cnts:
    # construct a mask from the contour
    mask = np.zeros(canvas.shape[:2], datatype="uint8")
    cv2.drawContours(mask, [c], -1, 255, -1)
    features = cv2.mean(canvas, mask=mask)[:3]
    data.append(features)
```

Lines 36–38 only threshold our image, resulting in a binary representation of the blobs, as shown in Figure 4-5.

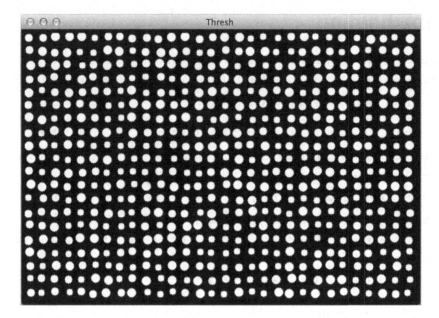

Figure 4-5. *The background and foreground are separated using thresholding to binarize our image*

On line 39, we capture the contours of the circles, create our data matrix (which will hold the features derived from each of the blobs), and begin looping through each of the individual contours on line 46, all from the aforementioned code. We create a mask for each contour and extract the average RGB values; we'll use these three averages to describe the circle's color.

Time to Cluster

Using the scikit-learn implementation of K-means, we can now do the actual clustering.

```
# cluster the hue features
clt = KMeans(n_clusters=3)
clt.fit(data)
cv2.imshow("Canvas", canvas)

# loop over the unique cluster identifiers
for i in np.unique(clt.labels_):
    # construct a mask for the current cluster
    mask = np.zeros(canvas.shape[:2], datatype="uint8")

    # loop over the indexes of the current cluster and
    sketch them
    for j in np.where(clt.labels_ == i)[0]:
        cv2.sketchContours(mask, [cnts[j]], -1, 255, -1)

    # show the output image for the cluster
    cv2.imshow("Cluster", cv2.bitwise_and(canvas, canvas,
    mask=mask))
    cv2.waitKey(0)
```

On line 54, we declare our K-means class and set the *n* clusters parameter to 3 to signify that we want three different clusters—one each for red, green, and blue. The fit technique is then used to cluster our data matrix.

We need a mechanism to retrieve the blobs that belong to each cluster now that we have our three clusters. As we already have the cnts variable holding the contours of our blobs, this method is a lot easier than it appears. To complete this example, all we need is a smart use of NumPy functions.

On line 59, we begin by looping through each of the unique cluster labels. Then, for each of the clusters, we use the np.where function to create a mask and collect the indices of all labels that correspond to the current cluster (line 64). We can utilize the indices to draw just the outlines that correspond to the current cluster now that we have them (line 65).

101

Revealing the Current Cluster

Finally, we'll use a bitwise operation to show just blobs that belong to the current cluster while concealing all others.

Simply use the following command to run our script:

```
$ python cluster_hues.py
```

Using the K-means method, we can observe that our blobs have been automatically sorted into the colors of red, green, and blue. We can see that just the shades of red have been segregated in Figure 4-6.

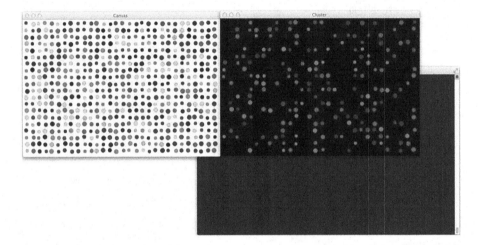

Figure 4-6. *The red blobs are being separated from the blues and greens*

Only hues of green are depicted in Figure 4-7.

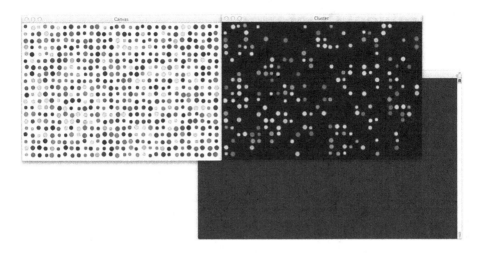

Figure 4-7. *Only the green circles are being extracted*

Lastly, we obtain just blue hues, as shown in Figure 4-8.

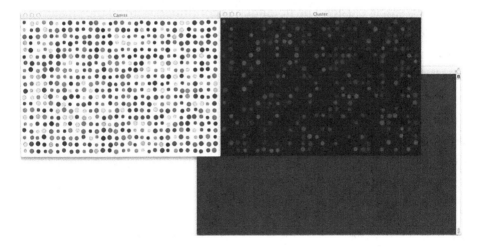

Figure 4-8. *Only looking at the "blue blob" cluster*

Now we will discuss another type of clustering algorithm.

Hierarchical Clustering

Hierarchical clustering is an iterative method of clustering data objects. There are two types.

- Agglomerative hierarchical algorithms, or a bottom-up approach

- Divisive hierarchical algorithms, or a top-down approach

Bottom-Up Approach

The bottom-up clustering method is called *agglomerative hierarchical clustering*. In this approach, each input object is considered as a separate cluster. In each iteration, an algorithm merges the two most similar clusters into only a single cluster. The operation is continued until all the clusters merge into a single cluster. The complexity of the algorithm is $O(n^3)$.

In the algorithm, a set of input objects, $I = \{I_1, I_2, ..., I_n\}$, is given. A set of ordered triples is <D,K,S>, where D is the threshold distance, K is the number of clusters, and S is the set of clusters.

Some variations of the algorithm might allow multiple clusters with identical distances to be merged into a single iteration. The following is the algorithm:

Input: $I = \{I_1, I_2, ..., I_n\}$
Output: O

```
for i = 1 to n do
        Ci ← {Ii};
end for
D ← 0;
K ← n;
S ← {C1,....., Cn};
O ← <d, k, S>;
```

repeat

      Dist ← CalcultedMinimumDistance(S);

      D ← ∞;

      For i = 1 to K-1 **do**

            For j = i+1 to K **do**

                  if Dist(i, j)< D **then**

                        D← Dist(i, j);

                        u ← i;

                        v ← j;

                end if

            end for

      end for

      K ← K-1;

      Cnew ← Cu ∪Cv;

      S ← S∪ Cnew -Cu - Cv;

      O ← O∪<D, K, S>

Until K = 1;

A Python example of hierarchical clustering is given later in the chapter.

Distance Between Clusters

In hierarchical clustering, calculating the distance between two clusters is a critical step. There are three methods to calculate this.

- Single linkage method

- Complete linkage method

- Average linkage method

Single Linkage Method

In the single linkage method, the distance between two clusters is the minimum distance of all distances between pairs of objects in two clusters. As the distance is the minimum, there will be a single pair of objects that has a less than equal distance between two clusters. So, the single linkage method may be given as follows:

$$\text{Dist}\left(C_i, C_j\right) = \min \text{dist}(X, Y)$$
$$X \in C_i, Y \in C_j$$

Complete Linkage Method

In the complete linkage method, the distance between two clusters is the maximum distance of all distances between pairs of objects in two clusters. The distance is the maximum, so all pairs of distances are less than or equal to the distance between two clusters. So, the complete linkage method can be given by the following:

$$\text{Dist}\left(C_i, C_j\right) = \max \text{dist}(X, Y)$$
$$X \in C_i, Y \in C_j$$

Average Linkage Method

The average linkage method is a compromise between the previous two linkage methods. It avoids the extremes of large or compact clusters. The distance between clusters C_i and C_j is defined by the following:

$$\text{Dist}\left(C_i, C_j\right) = \frac{\sum_{X \in Ci} \sum_{Y \in C_j} \text{dist}(X, Y)}{|C_i| \times |C_j|}$$

$|C_k|$ is the number of data objects in cluster C_k.

The centroid linkage method is similar to the average linkage method, but here the distance between the two clusters is actually the distance between the centroids. The centroid of cluster C_i is defined as follows:

$$X_c = (c_1, \ldots, c_m), \text{with}$$
$$c_j = 1/m \sum X_{kj},$$

X_{kj} is the j-th dimension of the k-th data object in cluster C_i.

Top-Down Approach

The top-down clustering method is also called the *divisive hierarchical clustering*. It the reverse of bottom-up clustering. It starts with a single cluster consisting of all input objects. After each iteration, it splits the cluster into two parts having the maximum distance. Here is the algorithm:

Input: I = {I1, I2, ... , In}

Output: O

```
D ← ∞;
K ← 1;
S ← {I₁,I₂ , ... , Iₙ};
0 ← <D, K, S >;
repeat
        X ← containing two data objects with the longest
            distance dist;
        Y ← ∅;
        S ← S - X;
        Xᵢ ← data object in A with maximum D (Xᵢ, X);
        X ← X − {Xᵢ};
        Y ← Y ∪ {Xᵢ};
        repeat
                for all data object Xⱼ in X do
```

$$e(j) \leftarrow\leftarrow \bar{D}\ (X_j, X)\ \bar{D}\ (X_j, Y);$$

        end for

        if$\exists$e(j) > 0 **then**

                $X_k \leftarrow\leftarrow$ data object in X with maximum e(j);

                $X \leftarrow X\ \{X_k\};$

                $Y \leftarrow\leftarrow Y \cup\cup \{X_k\};$

                split $\leftarrow\leftarrow$ TRUTH;

        else

                split$\leftarrow$ FALSE;

        end if

    until split == FALSE;

    D $\leftarrow$ dist;

    K $\leftarrow$ K+1;

    S $\leftarrow$ S$\cup\cup$ X $\cup$ Y

    O $\leftarrow$ O $\cup$ <D, K, S>;

Until K = n;

A dendrogram O is an output of any hierarchical clustering. Figure 4-9 illustrates a dendrogram.

Cluster Dendrogram

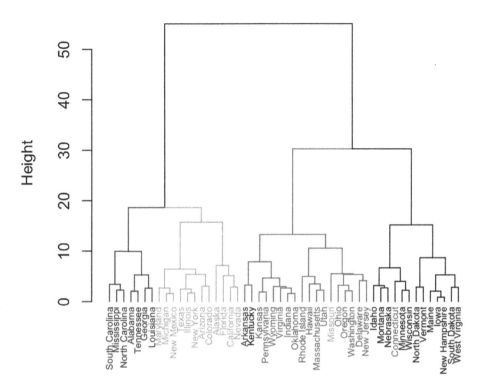

Figure 4-9. *A dendrogram*

To create a cluster from a dendrogram, you need a threshold of distance or similarity. An easy way to do this is to plot the distribution of distance or similarity and find the inflection point of the curve. For Gaussian distribution data, the inflection point is located at x = mean + n*std and x = mean – n*std, as shown in Figure 4-10.

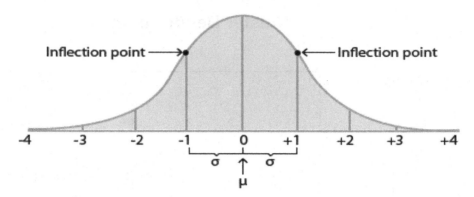

Figure 4-10. *The inflection point*

The following code creates a hierarchical cluster using Python. The input data file can be downloaded from https://canvas.uw.edu/courses/1546077/files/92110986.

```
import numpy as np
import pandas as pd
from sklearn.preprocessing import StandardScaler, normalize
from sklearn.decomposition import PCA
import scipy.cluster.hierarchy as hc
import matplotlib.pyplot as plt
from sklearn.cluster import AgglomerativeClustering

# Load the CSV file
df = pd.read_csv("book\\ch4\\CC_GENERAL.csv")

# preprocessing
df = df.drop('CUST_ID', axis = 1)
df.fillna(method ='ffill', inplace = True)

# Scale and normalize
ss = StandardScaler()
scaled_df = ss.fit_transform(df)
normalized_df= normalize(scaled_df)
```

```
# Reduce the dimensionality to 3
pca = PCA(n_components=3)
df_pca = pca.fit_transform(normalized_df)
df_pca = pd.DataFrame(df_pca)
df_pca.columns = ['P1', 'P2', 'P3']

# Dendogram plot
plt.figure(figsize =(10, 10))
plt.title('Visualising the data')
dendrogram = hc.dendrogram((hc.linkage(df_pca, method ='ward')))

# create 3 clusters using Agglomerative hierarchical clustering
hrc= AgglomerativeClustering(n_clusters = 3)
plt.figure(figsize =(10, 10))
plt.scatter(df_pca['P1'], df_pca['P2'], df_pca['P3'], c = hrc.
fit_predict(df_pca), cmap ='rainbow')
plt.title("Agglomerative Hierarchical Clusters - Scatter Plot",
fontsize=18)
plt.show()
```

Graph Theoretical Approach

The clustering problem can be mapped to a graph, where every node in the graph is an input data point. If the distance between two graphs is less than the threshold, then the corresponding nodes are connected. Now using the graph partition algorithm, you can cluster the graph. One industry example of clustering is in investment banking, where the cluster instruments depend on the correlation of their time series of price and performance trading of each cluster taken together. This is known as *basket trading* in algorithmic trading. So, by using the similarity measure, you can construct the graph where the nodes are instruments and the edges between the nodes indicate that the instruments are correlated. To create the basket, you need a set of instruments where all are correlated to each other. In a graph, this is a set of nodes or subgraphs where all the nodes in the subgraph are connected to each other. This kind of subgraph is known as a *clique*. Finding the clique of maximum size is an NPcomplete problem. People use heuristic solutions to solve this problem of clustering.

How Do You Know If the Clustering Result Is Good?

After applying the clustering algorithm, verifying the result as good or bad is a crucial step in cluster analysis. Three parameters are used to measure the quality of cluster, namely, the centroid, radius, and diameter.

$$\text{Centroid} = C_m = \frac{\sum_{i=1}^{N} t_{mi}}{N}$$

$$\text{Radius} = R_m = \sqrt{\frac{\sum_{i=1}^{N} \left(t_{mi} - C_m\right)^2}{N}}$$

$$\text{Diameter} = D_m = \sqrt{\frac{\sum_{i=1}^{N} \sum_{j=1}^{N} \left(t_{mi} - C_m\right)^2}{(N)(N-1)}}$$

If you consider the cluster as a circle in a space surrounding all member points in that cluster, then you can take the centroid as the center of the circle. Similarly, the radius and diameter of the cluster are the radius and diameter of the circle. Any cluster can be represented by using these three parameters. One measure of good clustering is that the distance between centers should be greater than the sum of radius.

The following code snippet from the previous script can be used to visualize the output where data points from the same cluster should be seen together.

```
# create 3 clusters using Agglomerative hierarchical clustering
hrc= AgglomerativeClustering(n_clusters = 3)
plt.figure(figsize =(10, 10))
plt.scatter(df_pca['P1'], df_pca['P2'], df_pca['P3'], c = hrc.
fit_predict(df_pca), cmap ='rainbow')
plt.title("Agglomerative Hierarchical Clusters - Scatter Plot",
fontsize=18)
plt.show()
```

Summary

Except for these chapters, every other chapter will contain data that we will use to label our model train. Then there's a fresh set of data that doesn't have a label. The model labels the data. This is called *supervised learning*.

However, there is no data label in this chapter. An algorithm groups or labels them based on their similarity. In a non-English-speaking country, children study English in school under the supervision of a teacher. However, after watching YouTube videos for two to three years, children begin to speak a modest amount of English without any instruction. This is called unsupervised learning.

CHAPTER 5

Deep Learning and Neural Networks

Neural networks, specifically known as *artificial neural networks* (ANNs), were developed by the inventor of one of the first neurocomputers, Dr. Robert Hecht-Nielsen. He defines a neural network as follows: "…a computing system made up of a number of simple, highly interconnected processing elements, which process information by their dynamic state response to external inputs."

Customarily, neutral networks are arranged in multiple layers. The layers consist of several interconnected nodes containing an activation function. The input layer, communicating to the hidden layers, delineates the patterns. The hidden layers are linked to an output layer.

Neural networks have many uses. As an example, you can cite the fact that in a passenger load prediction in the airline domain, passenger load in month t is heavily dependent on t-12 months of data rather on t-1 or t-2 data. Hence, the neural network normally produces a better result than the time-series model or even image classification. In a chatbot system, the memory network, which is actually a neural network of a bag of words of the previous conversation, is a popular approach. There are many ways to realize a neural network.

© Sayan Mukhopadhyay, Pratip Samanta 2023
S. Mukhopadhyay and P. Samanta, *Advanced Data Analytics Using Python*,
https://doi.org/10.1007/978-1-4842-8005-0_5

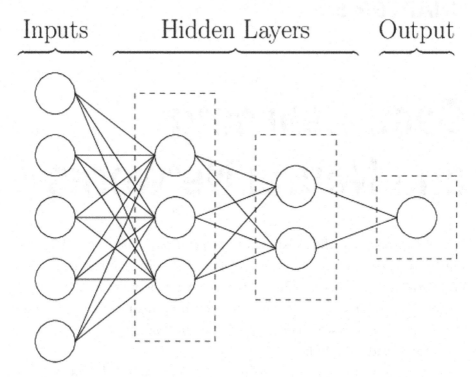

Figure 5-1. *Neural network architecture*

Backpropagation

Backpropagation, which usually substitutes an optimization method like gradient descent, is a common method of training artificial neural networks. The method computes the error in the outermost layer and backpropagates up to the input layer and then updates the weights as a function of that error, input, and learning rate. The final result is to minimize the error as far as possible.

Backpropagation Approach

Apply the input vector $X_p = (x_{p1}, x_{p2}, ..., x_{pN})^t$ to the input units.

Calculate the net input values to the hidden layer units.

$$net_{pj}^h = \sum_{i=1}^{N} \omega_{ji}^h x_{pi} + \theta_j^h$$

Calculate the outputs from the hidden layer.

$$i_{pj} = f_j^h \left(net_{pj}^h \right)$$

Calculate the net input values to each unit.

$$net_{pk}^o = \sum_{j=1}^{L} \omega_{kj}^o i_{pj} + \theta_k^o$$

Calculate the outputs.

$$o_{pk} = f_k^o \left(net_{pk}^o \right)$$

Calculate the error terms for the output units.

$$\delta_{pk}^o = \left(y_{pk} - o_{pk} \right) f_k^{o\prime} \left(net_{pk}^o \right)$$

Calculate the error terms for the hidden units.

$$\delta_{pj}^h = f_j^{h\prime} \left(net_{pj}^h \right) \sum_k \delta_{pk}^o \omega_{kj}^o$$

Update weights on the output layer.

$$\omega_{kj}^o (t+1) = \omega_{kj}^o (t) + \eta \delta_{pk}^o i_{pj}$$

Update weights on the hidden layer.

$$\omega_{ji}^h (t+1) = \omega_{ji}^h (t) + \eta \delta_{pj}^h x_i$$

Let's see some code:

```python
from numpy import exp, dot, array, random

class SimpleNN():
    def __init__(self):
        random.seed(2)
        self.weights = random.random((3, 1))

    # activation funtion
    def __sigmoid(self, x):
        return 1 / (1 + exp(-x))

    # derivative of the Sigmoid function.
    def __sigmoid_derivative(self, x):
        return x * (1 - x)

    # train the neural network and adjust weights
    def train(self, training_set_inputs, training_set_outputs,
    number_of_training_iterations):
        for iteration in range(number_of_training_iterations):
            output = self.predict(training_set_inputs)
            error = training_set_outputs - output
            adjustment = dot(training_set_inputs.T, error *
            self.__sigmoid_derivative(output))
            self.weights += adjustment

    # prediction
    def predict(self, inputs):
        return self.__sigmoid(dot(inputs, self.weights))

if __name__ == "__main__":

    neural_network = SimpleNN()

    print("Random starting weights: ")
```

```
print(neural_network.weights)

# The training set. We have 4 examples, each consisting of
3 input values
# and 1 output value.
training_set_inputs = array([[0, 0, 1], [1, 1, 1], [1, 0,
1], [0, 1, 1]])
training_set_outputs = array([[0, 1, 1, 0]]).T

neural_network.train(training_set_inputs, training_set_
outputs, 10000)

print("New weights after training: ")
print(neural_network.weights)

# Test the neural network with a new situation.
print(neural_network.predict(array([1, 0, 0])))
```

Other Algorithms

Many techniques are available to train neural networks besides
backpropagation. One of the methods is to use common optimization
algorithms such as gradient descent, the Adam optimizer, and so on. The
simple perception method is also frequently applied. Hebb's postulate
is another popular method. In Hebb's learning, instead of the error, the
product of the input and output goes as the feedback to correct the weight.

$$w_{ij}(t+1) = w_{ij}(t) + \eta y_j(t) x_i(t)$$

TensorFlow

TensorFlow is a popular deep learning library in Python. It is a Python wrapper on the original library. It supports parallelism on the CUDA-based GPU platform. The following code is an example of MNIST digit classification with TensorFlow:

```
import tensorflow as tf

# getting mnist data set
mnist = tf.keras.datasets.mnist

# loads data and splits into train and test set
(x_train, y_train), (x_test, y_test) = mnist.load_data()
x_train, x_test = x_train / 255.0, x_test / 255.0

# model
model = tf.keras.models.Sequential([
  tf.keras.layers.Flatten(input_shape=(28, 28)),
  tf.keras.layers.Dense(128, activation='relu'),
  tf.keras.layers.Dropout(0.2),
  tf.keras.layers.Dense(10, activation="softmax")
])

# loss funtion
loss_fn = tf.keras.losses.SparseCategoricalCrossentropy(fr
om_logits=True)

# model complie
model.compile(optimizer='adam',
              loss=loss_fn,
              metrics=['accuracy'])

# model fit
model.fit(x_train, y_train, epochs=5)
```

```
# model evaluation
model.evaluate(x_test,  y_test, verbose=2)
```

People nowadays do not use raw TensorFlow code. They use the wrapper of Keras, which is code that uses data from several family surveys to determine the risk of delivery.

You can find the data in the deep_learning_keras_1st_example folder in the Git repository of the book. Please refer to https://github.com/Apress/advanced-data-analytics-python-2e/tree/main/deep_learning_keras_1st_example.

```
# Importing modules
import pandas as pd
import numpy as np
import csv
from keras.layers.core import Dense, Activation, Dropout
from keras.layers.recurrent import LSTM
from keras.models import Sequential
import sys
from keras.preprocessing.sequence import TimeseriesGenerator
from sklearn.preprocessing import StandardScaler
from sklearn.metrics import classification_report, confusion_
matrix, ConfusionMatrixDisplay
import keras
import keras
from sklearn.decomposition import PCA
from sklearn.model_selection import train_test_split
from plot_keras_history import plot_history
import matplotlib.pyplot as plt
from sklearn.ensemble import RandomForestClassifier
from keras.utils import np_utils
import pandas as pd
```

```
PATH_TO_FOLDER = ''
df_delivery = pd.read_csv(PATH_TO_FOLDER+"delivery (1).csv")
print(df_delivery.head())
df_family_survey = pd.read_csv(PATH_TO_FOLDER+"family_survey
(1).csv")
print(df_family_survey.head())
df_merged = pd.merge(df_delivery, df_family_survey, how="left",
left_on=  ['hh_id'], right_on= ['hh_id'])
print(df_merged.columns)
print(df_merged.head())
print(df_merged.danger_signs_at_delivery.value_counts())

# Dropping unnecessary columns and typecasting

columns = ['delivery_id', 'patient_id', 'hh_id', 'delivery_
date_time_submitted']
df_merged.drop(columns, inplace=True, axis=1)
print(df_merged.columns)
df_merged = pd.get_dummies(df_merged, columns = ['facility_
delivery',  'first_visit_on_time', 'hand_washing_facilities',
'electricity', 'floor',
        'highest_education_achieved'])
print(df_merged.head())
print(df_merged.columns)
y = pd.Categorical(df_merged.danger_signs_at_delivery).codes
print(y[:10])

# Scaling and PCA

X = df_merged
X.fillna(0,inplace=True)
np.nan_to_num(X)
# scaled_X = X
Xscaler = StandardScaler()
```

```
Xscaler.fit(X)
scaled_X = Xscaler.transform(X)
np.nan_to_num(scaled_X)
pca = PCA(.95)
pca.fit(scaled_X)
scaled_X = pca.transform(scaled_X)
print(scaled_X.shape)

#Train, test set splitting

train_x, test_x, train_y, test_y = train_test_split( scaled_X,
np.array(y), test_size=1/7.0, random_state=0)

train_y=np_utils.to_categorical(train_y,num_classes=2)
test_y=np_utils.to_categorical(test_y,num_classes=2)
print("Shape of y_train",train_y.shape)
print("Shape of y_test",test_y.shape)

# Model architecture

model=Sequential()
model.add(Dense(500,input_dim=scaled_X.
shape[1],activation='relu'))
model.add(Dense(200,activation='relu'))
model.add(Dense(100,activation='relu'))
model.add(Dropout(0.1))
model.add(Dense(2,activation='sigmoid'))
model.compile(loss='binary_crossentropy',optimizer='adam',
metrics=['accuracy',keras.metrics.Recall()])
print(model.summary())
history = model.fit(train_x,train_y,validation_
data=(test_x,test_y),batch_size=500,epochs=4,verbose=1)

plot_history(history.history, path="standard.png")
plt.show()
```

```
prediction=model.predict(test_x)
y_label=np.argmax(test_y,axis=1)
predict_label=np.argmax(prediction,axis=1)

# accuracy=np.sum(y_label==predict_label)/length * 100
# print("Accuracy of the dataset",accuracy )
print(classification_report(y_label, predict_label))
```

Network Architecture and Regularization Techniques

Before moving on to the next section, not that as you raise the number of hidden layers in your network, the accuracy improves, but the application consumes more memory. Typically, people employ two to three hidden layers.

The Adam optimizer is commonly chosen because it combines gradient descent and stochastic gradient descent. People may add the coefficient of the resulting equation in the loss function because a big coefficient indicates overfitting. Another option is to use a dropout layer, which ignores a certain percentage of neurons chosen at random during time learning. For the classification problem, an entropy function, which is a measurement of the chaos of the system, should be used in the loss function. For binary and multiclass classification, there are multiple versions.

Note that the PCA presented in Chapter 3 is used here with a 95 percent variance. That is the standard.

Updatable Model and Transfer Learning

In deep learning, declaring a model trainable is all that is required to create an updatable machine learning model, as mentioned in Chapter 3. The following code is an example of an anomaly detection system. Every cloud service offers network packet information, and any instrument, such as Suricata, can send a security alert. The following preprocessor code processes data and marks the network packet as alert or nonalert depending on the alert type.

You can find the data in the frame_packet_june_14.csv file in the Git repository of the book. Please refer to the following links:

https://github.com/Apress/advanced-data-analytics-python-2e/blob/main/eve.json

https://github.com/Apress/advanced-data-analytics-python-2e/blob/main/frame_packet_June14.csv

```
# importing modules
import pandas as pd
import json
import gzip
import glob
import sys, codecs

#reading packet files and building data frame
PATH_TO_FOLDER = ""
# PATH_TO_FOLDER = "/home/ubuntu/AWS/"
df = pd.read_json(codecs.open(PATH_TO_FOLDER+'new-data\
sample-1.json','r','utf-8').read().replace("\\","/"),
lines=True)
print(df.columns)
#adding alert type and required typecasting
df['class_label'] = 0
df['alert_type'] = ""
# df['StartTime'] = pd.to_datetime(df['StartTime'])
df['src_port'] = df['src_port'].astype(int,errors = 'ignore')
df['dst_port'] = df['dst_port'].astype(int,errors = 'ignore')
df['source'] = df['source'].str.strip()
df['destination'] = df['destination'].str.strip()
print(df.head())
```

```
#for event txt file and update df if alert found
with open(PATH_TO_FOLDER+'new-data\eve.json','r') as fjson:
    for line in fjson:
        data = json.loads(line)#.decode('utf-8'))
        if data['event_type'] == 'alert':
            src_ip = data['src_ip'].strip()
            src_port = int(data['src_port'])
            dest_ip = data['dest_ip'].strip()
            dest_port = int(data['dest_port'])
            df.loc[(df.source == src_ip) & (df.destination ==
            dest_ip) & (df.src_port == src_port) & (df.dst_port
            == dest_port),['class_label','alert_type']] = [1,
            data['alert']['category']]

#saving into training file csv
df_label = df[df.class_label == 1]
print(df_label)
df.to_csv('frame_packet_June14.csv')
```

Here's the first iteration code (build the model and save it):

```
##  Filename : predict_type_packet.py
##  Purpose/Description : Main Classification Code
##  Author : Sayan Mukhopadhyay

''' module to predict packet type and generate classification
report '''
# importing modules
import pandas as pd
import numpy as np
import csv
from keras.layers.core import Dense, Activation, Dropout
from keras.layers.recurrent import LSTM
```

```python
from keras.models import Sequential
import sys
from keras.preprocessing.sequence import TimeseriesGenerator
from sklearn.preprocessing import StandardScaler
from sklearn.metrics import classification_report, confusion_
matrix, ConfusionMatrixDisplay
import keras
import keras
from sklearn.decomposition import PCA
from sklearn.model_selection import train_test_split
from plot_keras_history import plot_history
import matplotlib.pyplot as plt
from sklearn.ensemble import RandomForestClassifier
from keras.utils import np_utils

'''global variables/parameters'''
UNITS1 = 1000
UNITS2 = 500
UNITS3 = 300
EPOCHS = 4
DROPOUT_RATE = 0.2
BATCH_SIZE = 500

'''reading training csv'''
df = pd.read_csv('frame_packet_June14.csv', nrows=49000)

'''droppping unneccessary columns and typecasting '''
columns = ['id', 'conn_flag', 'pckt_info', 'TSval', 'TSecr',
'SLE', 'SRE']#, 'time']
df.drop(columns, inplace=True, axis=1)
columns2 = ['Unnamed: 0', 'class_label']
df.drop(columns2, inplace=True, axis=1)
print(len(df.columns))
```

```
df = df.replace('-',0.0)
df = pd.get_dummies(df, columns = ['src_port', 'dst_port',
'source', 'destination', 'protocol'])
print(len(df.columns))
print(df.columns)
df['time'] = pd.to_datetime(df['time'])
df = df.sort_values(by="time")
df.set_index('time', inplace=True)
print(df.head())
df.alert_type = df.alert_type.fillna('NonAlert')
df['alert_type'] = df['alert_type'].astype('category')

'''coverting categorical columns into codes'''
df['alert_type'] = pd.Categorical(df.alert_type).codes
y = df['alert_type'].tolist()
print(y[:10])
labels = df['alert_type'].unique()
print(labels)
count = len(df['alert_type'].unique())
print("alert_type count ",count)
columns3 = ['alert_type']
df.drop(columns3, inplace=True, axis=1)

'''standardscaler transformation and PCA, chossing important
columns'''
X = df
X.fillna(0,inplace=True)
np.nan_to_num(X)
# scaled_X = X
Xscaler = StandardScaler()
Xscaler.fit(X)
scaled_X = Xscaler.transform(X)
```

```
np.nan_to_num(scaled_X)
pca = PCA(400)
pca.fit(scaled_X)
scaled_X = pca.transform(scaled_X)
print(scaled_X.shape)

'''train, test set splitting label encoder'''
train_x, test_x, train_y, test_y = train_test_split( scaled_X,
np.array(y), test_size=1/7.0, random_state=0)

train_y=np_utils.to_categorical(train_y,num_classes=5)
test_y=np_utils.to_categorical(test_y,num_classes=5)
print("Shape of y_train",train_y.shape)
print("Shape of y_test",test_y.shape)

''' dnn model architecture '''
model=Sequential()
model.add(Dense(UNITS1,input_dim=scaled_X.
shape[1],activation='relu'))
model.add(Dense(UNITS2,activation='relu'))
model.add(Dense(UNITS3,activation='relu'))
model.add(Dropout(DROPOUT_RATE))
model.add(Dense(count,activation='softmax'))
model.compile(loss='categorical_crossentropy',optimizer='adam',
metrics=['accuracy', keras.metrics.Recall()])

''' model fit'''
history = model.fit(train_x,train_y,validation_
data=(test_x,test_y),batch_size=BATCH_
SIZE,epochs=EPOCHS,verbose=1)
model.save('my_model_400.h5')
```

```
prediction=model.predict(train_x)
length=len(prediction)
y_label=np.argmax(train_y,axis=1)
predict_label=np.argmax(prediction,axis=1)
print(classification_report(y_label, predict_label))
matrix = confusion_matrix(y_label, predict_label)
print(matrix)
```

Here is the all-iteration model (load the model, train it, and save it):

```
''' importing modules '''
import pandas as pd
import numpy as np
import csv
from keras.layers.core import Dense, Activation, Dropout
from keras.layers.recurrent import LSTM
from keras.models import Sequential
import sys
from keras.preprocessing.sequence import TimeseriesGenerator
from sklearn.preprocessing import StandardScaler
from sklearn.metrics import classification_report, confusion_
matrix, ConfusionMatrixDisplay
import keras
import keras
from sklearn.decomposition import PCA
from sklearn.model_selection import train_test_split
from plot_keras_history import plot_history
import matplotlib.pyplot as plt
from sklearn.ensemble import RandomForestClassifier
from keras.utils import np_utils
from keras.models import load_model

'''global variables/parameters'''
```

```python
'''reading training csv'''
df = pd.read_csv(' frame_packet_June14.csv')#, nrows=5000)
cols = ['time_x', 'source_x', 'destination_x', 'protocol_x',
'length_x', 'src_port',
        'dst_port_x', 'Seq_x', 'Ack_x', 'Win_x', 'Len_x',
'MSS_x', 'WS_x', 'SACK_PERM_x',
        'alert_type']
df = df[cols]
print(df.head())

'''droppping unneccessary columns and typecasting '''
df = df.replace('-',0.0)
df = pd.get_dummies(df, columns = ['src_port', 'dst_port_x',
'source_x', 'destination_x', 'protocol_x'])
print(len(df.columns))
print(df.columns)
print(df.loc[0,'time_x'])
df['time_x'] = pd.to_datetime(df['time_x'])
df = df.sort_values(by="time_x")
df.set_index('time_x', inplace=True)
df.alert_type = df.alert_type.fillna('NonAlert')
df['alert_type'] = df['alert_type'].astype('category')

'''coverting categorical columns into codes'''
codes = pd.Categorical(df.alert_type).codes
y = codes
print(y[:10])
count = len(df['alert_type'].unique())
print("alert_type count ",count)
columns3 = ['alert_type']
df.drop(columns3, inplace=True, axis=1)
```

```
'''standardscaler transformation and PCA, chossing important
columns'''
X = df
X.fillna(0,inplace=True)
np.nan_to_num(X)
Xscaler = StandardScaler()
Xscaler.fit(X)
scaled_X = Xscaler.transform(X)
np.nan_to_num(scaled_X)
pca = PCA(400)
pca.fit(scaled_X)
scaled_X = pca.transform(scaled_X)
print(scaled_X.shape)

'''train, test set splitting and timeseries generator'''
train_x, test_x, train_y, test_y = train_test_split( scaled_X,
np.array(y), test_size=1/7.0, random_state=0)

train_y=np_utils.to_categorical(train_y,num_classes=5)
test_y=np_utils.to_categorical(test_y,num_classes=5)
print("Shape of y_train",train_y.shape)
print("Shape of y_test",test_y.shape)

model = load_model('my_model_400.h5')
for i in range(3):
    model.layers[i].trainable = True
history = model.fit(train_x,train_y,validation_
data=(test_x,test_y),batch_size=500,epochs=4,verbose=1)
plot_history(history.history, path="standard.png")
plt.show()

'''model accuracy'''
prediction=model.predict(test_x)
y_label=np.argmax(test_y,axis=1)
```

```
predict_label=np.argmax(prediction,axis=1)
print(classification_report(y_label, predict_label))
labels = [i for i in range(0,count)]

matrix = confusion_matrix(y_label, predict_label)
print(matrix)

prediction=model.predict(train_x)
length=len(prediction)
y_label=np.argmax(train_y,axis=1)
predict_label=np.argmax(prediction,axis=1)
print(classification_report(y_label, predict_label))
matrix = confusion_matrix(y_label, predict_label)
print(matrix)
```

Note that PCA is not initialized with a 95 percent variance since we need the same number of parameters in each iteration; thus, we chose 15 as the number of parameters as the network packet had a total of 22 input parameters. PCA is not an updatable model. The development of an updatable feature selection model is still a research area.

This type of system is useful for a TV channel that wants to predict whether a user would churn. As they are not addicted to the channel, *churn* is defined as a user who does not renew their plan within 24 hours of it expiring. However, since hackers are continually trying new things, the network intrusion detection model always has new types of alerts. As a result, the system should cluster the data first, with each group representing a different type of alert. We have a classifier model for each category that determines if the alert is of that type.

Recurrent Neural Network

A *recurrent neural network* is an extremely popular kind of network where the output of the previous step goes to the feedback or is input to the hidden layer. It is an extremely useful solution for a problem like a sequence leveling algorithm or time-series prediction. One of the more popular applications of the sequence leveling algorithm is in an autocomplete feature of a search engine.

LSTM

In an RNN, the network takes feedback from past.$X_{(t)} = K \times X_{(t-1)} = K^2 \times X_{(t-2)} = K^N \times X_{(t-N)}$. Now, if K > 1, then K^N is very large; otherwise, if K < 1, then K^N is very small. To avoid this problem, network programmatically forgets some of its past state. LSTM does this.

This way, it can remember values over arbitrary intervals. LSTM works very well to classify, process, and predict time series given time lags of unknown duration. Relative insensitivity to gap length gives an advantage to LSTM over alternative RNNs, hidden Markov models, and other sequence learning methods.

RNN and HMM rely on the hidden state before emission/sequence. If we want to predict the sequence after 500 intervals instead of 5, LSTM can remember the states and predict properly.

To simplify this, suppose a person forgets old memories and remembers only recent things. But they need those memories from the past to perform some tasks in the future. This is the problem with traditional RNNs. Also, there is another person who remembers the important memories from the past along with the recent ones and deletes the useless memories from the past. This way, they can use that information to carry out the task more efficiently. This is the case with LSTM.

Each LSTM cell has three inputs, $h_{\{t-1\}}$, $c_{\{t-1\}}$, and x_t, and two outputs, h_t and c_t. For a given time t, h_t is the hidden state, c_t is the cell state or memory, and x_t is the current data point or input. The first sigmoid layer has two inputs, $h_{\{t-1\}}$ and x_t, where $h_{\{t-1\}}$ is the hidden state of the previous cell. It is known as the *forget gate* as its output selects the amount of information of the previous cell to be included. The output is a number in [0,1], which is multiplied (pointwise) with the previous cell state $c_{\{t-1\}}$.

The network packet is a sequence, so the RNN can readily be applied to an anomaly detection system. One difference is that we store data in a cloud database and sort it by IP address and ports. This sorting is crucial since without a suitable sequence, you will not get a good result. The autocorrelation function, which is discussed in Chapter 3, can be used to determine whether the sequence is correct.

```
# importing modules
import pandas as pd
import numpy as np
import csv
from keras.layers.core import Dense, Activation, Dropout
from keras.layers.recurrent import LSTM
from keras.models import Sequential
import sys
from keras.preprocessing.sequence import TimeseriesGenerator
from sklearn.preprocessing import StandardScaler
from sklearn.metrics import classification_report, confusion_
matrix, ConfusionMatrixDisplay
import keras
import keras
from sklearn.decomposition import PCA
from sklearn.model_selection import train_test_split
from plot_keras_history import plot_history
import matplotlib.pyplot as plt

sys.path.append('../resource')
from MSSqlDb import MSSqlDbWrapper

mssql_instance = MSSqlDbWrapper("../config/config1.txt")
con = mssql_instance.get_connect()
df = pd.read_sql("SELECT * from packet_table order by IP,
PORT;",con)
```

```python
#global variables/parameters
LSTM_UNITS1 = 150
LSTM_UNITS2 = 100
LSTM_UNITS3 = 50
EPOCHS = 3
DROPOUT_RATE = 0.2
TIMESERIESLEN = 50
#reading training csv

df = pd.get_dummies(df, columns = ['src_port', 'dst_port',
'source', 'destination', 'protocol'])
df['time'] = pd.to_datetime(df['time'])
df = df.sort_values(by="time")
df.set_index('time', inplace=True)
print(df.head())
df['alert_type'] = df['alert_type'].astype('category')

#coverting categorical columns into codes
df['alert_type'] = pd.Categorical(df.alert_type).codes
y = df['alert_type'].tolist()
labels = df['alert_type'].unique()
count = len(df['alert_type'].unique())
print("alert_type count ",count)
columns3 = ['alert_type']
df.drop(columns3, inplace=True, axis=1)

#standardscaler transformation and PCA, chossing
important columns
X = df
X.fillna(0,inplace=True)
np.nan_to_num(X)
Xscaler = StandardScaler()
Xscaler.fit(X)
```

```
scaled_X = Xscaler.transform(X)
np.nan_to_num(scaled_X)

#changing format of class labels
y_intermediate = [[0 for i in range(count)] for j in
range(len(y))]
for i in range(len(y)):
    y_intermediate[i][y[i]] = 1
y_final = np.array(y_intermediate)
# sys.exit()

#train, test set splitting and timeseries generator
train_x, test_x, train_y, test_y = train_test_split( scaled_X,
y_final, test_size=1/7.0, random_state=0)
train_generator = TimeseriesGenerator(train_x,train_y,
length=TIMESERIESLEN)
test_generator = TimeseriesGenerator(test_x, test_y,
length=TIMESERIESLEN)
print(len(test_generator))
for i in range(len(test_generator)):
  x, y = test_generator[i]
  print(len(y))
print(len(test_y))
print("train generator")
for i in range(len(train_generator)):
  x, y = train_generator[i]
  print(len(y))
print(len(train_y))

#LSTM model train
model =  Sequential()
model.add(LSTM(LSTM_UNITS1, activation='relu', input_
shape=(TIMESERIESLEN,scaled_X.shape[1]),return_sequences=True))
```

```
model.add(Dropout(DROPOUT_RATE))
model.add(LSTM(LSTM_UNITS2,return_sequences=True))
model.add(Dropout(DROPOUT_RATE))
model.add(LSTM(LSTM_UNITS3,return_sequences=False))
model.add(Dropout(DROPOUT_RATE))
model.add(Dense(count, activation='softmax'))
model.compile(optimizer='adam',loss='categorical_
crossentropy',metrics=[keras.metrics.Recall(), 'accuracy'])
history= model.fit_generator(train_generator,epochs=EPOCHS)

#model accuracy
plot_history(history.history, path="standard.png")
plt.show()
scores = model.evaluate(test_generator)
print("Final Score")
print(scores)
```

Reinforcement Learning

We'll talk about reinforcement learning in this section. Learning from feedback is referred to as *reinforcement learning*. Reinforcement learning is one of three main types of machine learning approach alongside supervised and unsupervised machine learning. It's used to learn models by performing specific tasks in a given environment. The program interacts with its surroundings and performs actions to move between different states. Actions are then either positively or negatively considered through reward or penalty. Successful actions are reinforced, and unsuccessful actions are penalized. A model will go through many different iterations to find the best possible sequence of actions to achieve a given goal. The following is the algorithm behind it.

TD0

Algorithm The TD(0) tabular algorithm is implemented by this function. After each transition, this function must be called.

function TD0(*X,R,Y,V*):

- *X* = Last state

- *Y* = Next State

- *R* = Instant reward connected with this transition

- *V* = Array of estimated value

$$\delta = R + \gamma \left[V(Y) - V(X) \right]$$

$$V[Y] = V[X] + \alpha \cdot \delta$$

where α is step size.

return *V*

Please pass the following file in a command prompt while running this function:

https://github.com/Apress/advanced-data-analytics-python-2e/blob/main/reinforcement_learning_td0_reduce_dat.csv

Here is an example of four-step TD0 with alpha and gamma 1:

```
import pandas as pd
import numpy as np
import sys
from sklearn.model_selection import train_test_split
import tensorflow as tf
from keras.models import Sequential
from keras.layers.core import Dense
from scipy.stats.stats import pearsonr
from math import sqrt
```

```python
if len(sys.argv) < 3:
     print("Usage is")
     print("python assignment.py    <input file
path>    <output file path>    <No of split>")
     exit(0)

#Read the input data
df = pd.read_csv(sys.argv[1])

split = int(sys.argv[3])

out = open(sys.argv[2],'w')

final_error = []

size = int(df.shape[0]/split)

matched = 0
relaxed_matched = 0
count = 0
square_sum = 0
sum = 0

#Run the code for each split of input
for i in range(split):
    X = df.loc[i*size: (i+1)*size]
    X = X.astype(float)
    #Fill the missing values by the average of the column
    X.fillna(X.mean(), inplace=True)
    y = X['y']
    X.drop('y', inplace=True, axis=1)

    X_back = X

    X = X.as_matrix()
    y = y.as_matrix()
```

```
#split the data in test and training sample
X_train, X_test, y_train, y_test = train_test_split(X, y,
test_size= 0.4, random_state=42)

#normalize the data
while(1):
    flag = True
    for i in range(X_train.shape[1]):
        if X_train[:,i].std() != 0:
            X_train[:,i] = (X_train[:,i]- X_train[:,i].
            mean())/X_train[:,i].std()
            X_test[:,i] = (X_test[:,i]- X_test[:,i].
            mean())/X_test[:,i].std()
        else:
            X_train = np.delete(X_train,i,1)
            X_test = np.delete(X_test,i,1)
            flag = False
            break
    if flag:
        break

av = y_train.mean()
st = y_train.std()
y_train = (y_train- y_train.mean())/y_train.std()

index = []
i1 = 0
processed = 0

#select the columns which is correlated with y
while(1):
    flag = True
    for i in range(X_train.shape[1]):
```

```
            if i > processed :
                i1 = i1 + 1
                corr = pearsonr(X_train[:,i], y_train)
                PEr= .674 * (1- corr[0]*corr[0])/ (len(X_
                train[:,i])**(1/2.0))
                if abs(corr[0]) < PEr:
                    X_train = np.delete(X_train,i,1)
                    X_test = np.delete(X_test,i,1)
                    index.append(X_back.columns[i1-1])
                    processed = i - 1
                    flag = False
                    break
        if flag:
            break
    #drop the columns which is correlated with other
    input column
    while(1):
        flag = True
        for i in range(X_train.shape[1]):
            for j in range(i+1,X_train.shape[1]-1):
                corr = pearsonr(X_train[:,i], X_train[:,j])
                PEr= .674 * (1- corr[0]*corr[0])/ (len(X_
                train[:,i])**(1/2.0))
                if abs(corr[0]) > 6*PEr:
                    X_train = np.delete(X_train,j,1)
                    X_test = np.delete(X_test,j,1)
                    flag = False
                    break
            break
        if flag:
            break
```

```
#build the model to predict the y
learning_rate = 0.0001

model = Sequential([
    Dense(64, activation=tf.nn.relu, input_shape=[X_train.
    shape[1]]),
    Dense(64, activation=tf.nn.relu),
    Dense(1)
])

optimizer = tf.train.RMSPropOptimizer(learning_rate)

model.compile(loss='mse',
              optimizer=optimizer,
              metrics=['mae', 'mse'])

model.fit(
X_train, y_train,
epochs=int(X_train.shape[1]/2), validation_split = 0.2,
verbose=0)

predict = model.predict(X_train)

#build the model to predict the error in prediction
error = []
for i in range(len(predict)):
    error.append(y_train[i] - predict[i][0])

error = np.array(error)

model_e = Sequential([
    Dense(64, activation=tf.nn.relu, input_shape=[X_train.
    shape[1]]),
    Dense(64, activation=tf.nn.relu),
    Dense(1)
])
```

```python
model_e.compile(loss='mse',
                optimizer=optimizer,
                metrics=['mae', 'mse'])

model_e.fit(
    X_train, error,
epochs=int(X_train.shape[1]/2), validation_split = 0.2,
verbose=0)

#predict the test data using the trained model
predict = model.predict(X_test)

err_p = model_e.predict(X_test)

predict = predict + err_p

predict = predict*st + av

for i in range(len(predict)):
    error = y_test[i] - predict[i][0]
    if abs(error) <= 3:
        matched = matched + 1
    if abs(error/y_test[i]) <= 0.1:
        relaxed_matched = relaxed_matched + 1
    square_sum = square_sum + error*error
    sum = sum + error
    count = count + 1
out.write("RMSE="+str(sqrt(square_sum/count))+'\n')
out.write("matched count="+ str(matched) +'\t Total count=' +
str(count) +'\n')

out.write("ME="+str(sqrt(abs(sum)/count))+'\n')
out.write("relaxed matched count="+ str(relaxed_matched) +'\t
Total count=' + str(count) +'\n')
```

```
out.close()

print("RMSE=",str(sqrt(square_sum/count)),'\n')
print("matched count=", str(matched),'\t', "Total count=",
str(count),'\n')

print("ME=",str(sqrt(abs(sum)/count)),'\n')
print("relaxed matched count=", str(relaxed_matched),'\t',
"Total count=", str(count),'\n')
```

You can use this method to boost accuracy in any regression problem by predicting error, but it's a little trickier for classification.

TDλ

Algorithm This function uses replacing traces to perform the tabular TD(λ) algorithm. After each transition, this function must be called.
    function TDLambda (X, R, Y, V, z)

- X= Last state

- Y= Next state

- R= Instant reward connected with this transition

- V = Array of estimated value

- z= Array of eligibility traces

$$\delta = R + \gamma \cdot V[Y] - V[X]$$

for all $x \in \mathcal{X}$ do:

$$z[x] = \gamma \cdot \lambda \cdot z[x]$$

```
if X = x then
  z[x] = 1
end if
V[X] = V[x] + α · δ · z[x]
end for
return (V, z)
```

Example of Dialectic Learning

An algorithmic trader now wants to divide stock prices into three goal categories: same, up, and down. The class same denotes that the stock's price has remained unchanged. The class up denotes that the stock's price is going up. The class down denotes that the stock's price is decreasing.

Ninety-seven percent of the data is classified as the class same. The time series is about 20,000 points long. For this type of problem, most people use biased sampling. We, on the other hand, did things differently. We divide the data into batches of 1,000 points and train the model with 1,000 data points to predict the following 100 in each iteration. In Keras, we use SoftMax regression with RNN with a sequence length of 100. We now calculate the probability of being up, down, or the same for each iteration. We also compute the probability distribution's mean and standard deviation. We now use the following formula to determine the score for each class:

```
inc = (prob_increasing[j] - increasing_mean + k_inc*
increasing_std)
dec = (prob_decreasing[j] - decreasing_mean + k_dec*
decreasing_std)
same = (prob_same[j] - same_mean + k_same*same_std)
```

where K_inc, K_dec, and K_same are the constants initialized as 1.

Then we classify the data using the following logic:

```
if same > 0:
    pr_status = 0
else:
    if inc > dec:
        pr_status = 100
    else:
        pr_status = -100
```

So, most of the data is classified in same and the rest is in the other class.

Then we calculate the following parameters in each iteration:

```
if acc_status == 0:
    if pr_status == 0:
        wrong_count_pos_same = wrong_count_pos_same + 1
    total_count_acc_same = total_count_acc_same + 1
else:
    if pr_status != 0:
        wrong_count_neg_same = wrong_count_neg_same + 1
    total_count_acc_not_same = total_count_acc_not_same + 1
```

where

- wrong_count_pos_same = Count of points wrongly classified in the same class

- total_count_acc_same = Count of points actually belonging to the same class

- wrong_count_neg_same = Count of points wrongly not classified in the same class

- total_count_acc_not_same = Count of points actually not in the class same

The same parameters are calculated for the up and down classes considering the points are not in the same class, and then after each iteration, the constants are adjusted with the following logic:

```
if total_count_acc_same != 0:
    if wrong_count_neg_same/total_count_acc_same > .5:
        k_same = 1.2 * k_same
    if total_count_acc_not_same !=0:
        if wrong_count_pos_same/total_count_acc_not_
        same > .5:
        k_same = 0.9 * k_same
```

So, a dialectic was created between all candidate classes, and the points were adjusted depending on local trends of 100 points on top of the prediction based on the previous 1,000 points. This pattern improves the accuracy of classification with any model in stock point prediction. The original code is given next.

Please copy the CSV files from https://github.com/Apress/advanced-data-analytics-python-2e.

```
''''

#import matplotlib.pyplot as plt
import numpy as np
import csv
from keras.layers.core import Dense, Activation, Dropout
from keras.layers.recurrent import LSTM
from keras.models import Sequential
import sys

def read_data(path_to_dataset, path_to_target,
              sequence_length=50,
              ratio=1.0):
```

```python
    max_values = ratio * 2049280

    with open(path_to_dataset) as f:
        data = csv.reader(f, delimiter=",")
        power = []
        nb_of_values = 0

        for line in data:
            try:
                power.append([float(line[1]),float(line[4]),
                float(line[7])])
                nb_of_values += 1
            except ValueError:
                pass
            if nb_of_values >= max_values:
                break

    with open(path_to_target) as f:
        data = csv.reader(f, delimiter=",")
        target = []
        nb_of_values = 0
        for line in data:
            try:
                target.append(float(line[0].strip()))
                nb_of_values += 1
            except ValueError:
                pass
            if nb_of_values >= max_values:
                break
    return power, target

def create_matrix(y_train):
    y = [[0 for i in range(3)] for j in range(len(y_train))]
```

```python
    for i in range(len(y_train)):
        if y_train[i] == -100:
            y[i][0] = 1
        else:
            if y_train[i] == 100:
                y[i][1] = 1
            else:
                if y_train[i] == 0:
                    y[i][2] = 1
    return y

def process_data(power, target, sequence_length):
    result = []

    for index in range(len(power) - sequence_length-1):
        result.append(power[index: index + sequence_length])
    result = np.array(result)

    #print(result.shape)

    row = int(round(0.9 * result.shape[0]))

    X_train = result[:row, :]

    #X_train = train[:, :-1]

    y_train = np.array(create_matrix(target))
    #print(y_train.shape)
    X_test = result[row:, :]
    y_test = y_train[row:]
    #print(y_test.shape)
    y_train = y_train[:row]
    #print(y_train.shape)
```

```
    #print(X_train.shape)

    X_train = np.reshape(X_train, (X_train.shape[0], X_train.
    shape[1], 3))
    X_test = np.reshape(X_test, (X_test.shape[0], X_test.
    shape[1], 3))

    return [X_train, y_train, X_test, y_test]
def build_model():
    model = Sequential()
    layers = [3, 100, 50, 3]

    model.add(LSTM(
        layers[1],
        input_shape=(None, layers[0]),
        return_sequences=True))
    model.add(Dropout(0.2))

    model.add(LSTM(
        layers[2],
        return_sequences=False))
    model.add(Dropout(0.2))

    model.add(Dense(
        layers[3]))
    model.add(Activation('softmax'))

    model.compile(loss="categorical_crossentropy",
    optimizer="adam")

    return model
def run_network(data=None, target=None):
    epochs = 2
```

```
    ratio = 0.5
    sequence_length = 50

    X_train, y_train, X_test, y_test = process_data(
        data, target, sequence_length)

    model = build_model()

    try:
        model.fit(
            X_train, y_train,
            batch_size=512, nb_epoch=epochs, validation_
            split=0.05, verbose=0)
        predicted = model.predict(X_test)
    except KeyboardInterrupt:
        exit(0)

    return y_test, predicted
def convert(x):
    if x[0] == 1:
        return -100
    if x[1] == 1:
        return 100
    if x[2] == 1:
        return 0

if __name__ == '__main__':
    path_to_dataset = ' dialectic_learnign_data.csv '
    path_to_target = ' dialectic_learning_label.csv '
    data, target = read_data(path_to_dataset, path_to_target)
    k_inc = 1
    k_dec = 1
    k_same = 1
```

```
for i in range(0,len(data)-1000,89):
    d = data[i:i+1001]
    t = target[i:i+1001]
    y_test, predicted = run_network(d,t)

    prob_increasing = predicted[:,1]
    increasing_mean = prob_increasing.mean()
    increasing_std = prob_increasing.std()
    prob_decreasing = predicted[:,0]
    decreasing_mean = prob_decreasing.mean()
    decreasing_std = prob_decreasing.std()
    prob_same = predicted[:,2]
    same_mean = prob_same.mean()
    same_std = prob_same.std()
    wrong_count_pos_same = 0
    total_count_acc_not_same = 0
    wrong_count_neg_same = 0
    total_count_acc_same = 0
    wrong_count_pos_up = 0
    total_count_acc_not_up = 0
    wrong_count_neg_up = 0
    total_count_acc_up = 0
    wrong_count_pos_down = 0
    total_count_acc_not_down = 0
    wrong_count_neg_down = 0
    total_count_acc_down = 0
    for j in range(len(predicted)-1):
        inc = (prob_increasing[j] - increasing_mean + k_
        inc*increasing_std)
        dec = (prob_decreasing[j] - decreasing_mean + k_
        dec*decreasing_std)
        same = (prob_same[j] - same_mean +  k_
        same*same_std)
```

```
acc_status = convert(y_test[j])
if same > 0:
    pr_status = 0
else:
    if inc > dec:
        pr_status = 100
    else:
        pr_status = -100

if acc_status == 0:
    if pr_status == 0:
        wrong_count_pos_same = wrong_count_pos_
        same + 1
    total_count_acc_same = total_count_acc_same + 1
else:
    if pr_status != 0:
        wrong_count_neg_same = wrong_count_neg_
        same + 1
    total_count_acc_not_same = total_count_acc_not_
    same + 1
    if acc_status == 100:
        if pr_status != 100:
            wrong_count_pos_up = wrong_count_
            pos_up + 1
        total_count_acc_up = total_count_acc_up + 1
    else:
        if pr_status == 100:
            wrong_count_neg_up = wrong_count_
            neg_up + 1
        total_count_acc_not_up = total_count_acc_
        not_up + 1
```

```
        if acc_status == -100:
            if pr_status != -100:
                wrong_count_pos_down = wrong_count_pos_
                down + 1
            total_count_acc_down = total_count_acc_
            down + 1
        else:
            if pr_status == -100:
                wrong_count_neg_down = wrong_count_neg_
                down + 1
            total_count_acc_not_down = total_count_acc_
            not_down + 1

    print(acc_status,',', pr_status)

if total_count_acc_same != 0:
    if wrong_count_neg_same/total_count_acc_same > .5:
        k_same = 1.2 * k_same
if total_count_acc_not_same !=0:
    if wrong_count_pos_same/total_count_acc_not_
    same > .5:
        k_same = 0.9 * k_same

if total_count_acc_up != 0:
    if wrong_count_neg_up/total_count_acc_up > .5:
        k_inc = 1.2 * k_inc
if total_count_acc_not_up !=0:
    if wrong_count_pos_up/total_count_acc_not_up > .5:
        k_inc = 0.9 * k_inc

if total_count_acc_down != 0:
    if wrong_count_neg_down/total_count_acc_down > .5:
        k_dec = 1.2 * k_dec
```

```
if total_count_acc_not_down !=0:
    if wrong_count_pos_down/total_count_acc_not_
    down > .5:
        k_dec = 0.9 * k_dec
```

You can find the data in the `dialectic_learning_data.csv` and `dialectic_learning_label.csv` files in the Git repository of the book.

Convolution Neural Networks

Now we will discuss another kind of neural network: convolution neural networks.

Each neuron in the input layer is linked to each output neuron in the following layer, forming a fully connected (FC) layer in classic feedforward neural networks (like the ones we studied previously in this book). In CNNs, however, FC layers are not used until the very final layer (or final few layers of a network).

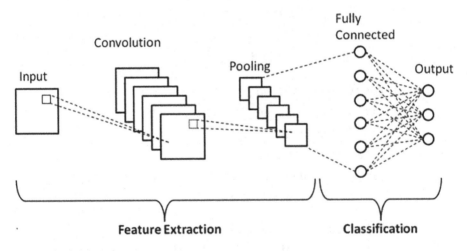

Figure 5-2. *CNN architecture*

Instead, we compute the output using convolutional filters applied to the input layer. When you use these convolutions, you get local connections, which means that every part of the input is linked to a portion of the output (we'll explain this later in the chapter). In a CNN, each layer applies a separate set of filters, usually hundreds or thousands, and then merges the results.

During training, a CNN learns the values for these filters automatically. When it comes to picture categorization, our CNN could learn to do the following:

- In the first layer, detect edges using raw pixel data.

- In the second layer, use these edges to recognize shapes (i.e., blobs).

- In the top layers of the network, use these shapes to detect higher-level characteristics like face structures.

- The last layer is a classifier, which makes predictions about the contents of the picture based on these higher-level characteristics.

An RNN of a CNN is a suitable choice if you have a sequence of images, such as a video. We classify a sequence of images in the following code:

```
import numpy as np
from keras.models import Sequential
from keras.layers import Conv2D, MaxPooling2D, Dense, Flatten
from keras.utils import to_categorical
import tensorflow as tf

mnist = tf.keras.datasets.mnist

# loads data and splits into train and test set
(x_train, y_train), (x_test, y_test) = mnist.load_data()
x_train, x_test = x_train / 255.0, x_test / 255.0
```

```python
# # Reshape the images.
train_images = np.expand_dims(x_train, axis=3)
test_images = np.expand_dims(x_test, axis=3)

num_filters = 4
filter_size = 2
pool_size = 2

# Build the model.
model = Sequential([
  Conv2D(num_filters, filter_size, input_shape=(28, 28, 1)),
  MaxPooling2D(pool_size=pool_size),
  Flatten(),
  Dense(10, activation='softmax'),
])

# Compile the model.
model.compile(
  'adam',
  loss='categorical_crossentropy',
  metrics=['accuracy'],
)

# Train the model.
model.fit(
  train_images,
  to_categorical(y_train),
  epochs=3,
  validation_data=(test_images, to_categorical(y_test)),
)
```

```
# Predict on the first 10 test images.
predictions = model.predict(test_images[:10])
print(np.argmax(predictions, axis=1))

# Check our predictions against the ground truths.
print(x_test[:10])
```

Summary

This chapter is the heart of this book. We discussed neural networks such as RNN and CNN with reinforcement learning using real examples. In the next chapter, we will discuss some classical statistical methods to analyze time-series data, and the last chapter is all about how to scale your analytic application.

CHAPTER 6

Time Series

A *time series* is a series of data points arranged chronologically. Most commonly, the time points are equally spaced. A few examples are the passenger loads of an airline recorded each month for the past two years or the price of an instrument in the share market recorded each day for the last year. The primary aim of time-series analysis is to predict the future value of a parameter based on its past data.

Classification of Variation

Traditionally, time-series analysis divides the variation into three major components, namely, trends, seasonal variations, and other cyclic changes. The variation that remains is attributed to "irregular" fluctuations or error term. This approach is particularly valuable when the variation is mostly comprised of trends and seasonality.

Analyzing a Series Containing a Trend

A *trend* is a change in the mean level that is long-term in nature. For example, if you have a series like 2, 4, 6, 8 and someone asks you for the next value, the obvious answer is 10. You can justify your answer by fitting a line to the data using the simple least square estimation or any other regression method. A trend can also be nonlinear. Figure 6-1 shows an example of a time series with trends.

© Sayan Mukhopadhyay, Pratip Samanta 2023
S. Mukhopadhyay and P. Samanta, *Advanced Data Analytics Using Python*,
https://doi.org/10.1007/978-1-4842-8005-8_6

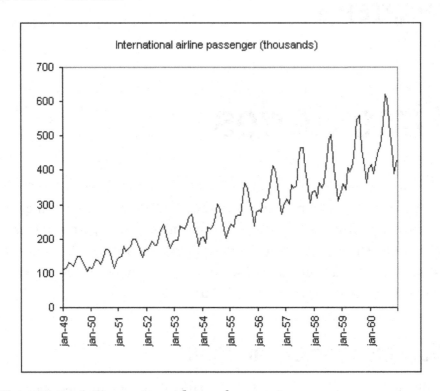

Figure 6-1. *A time series with trends*

The simplest type of time series is the familiar "linear trend plus noise" for which the observation at time *t* is a random variable X_t, as follows:

$$X_t = \alpha + \beta t + \varepsilon_t$$

Here, α, β are constants, and ε_t denotes a random error term with a mean of 0. The average level at time *t* is given by $m_t = (\alpha + \beta t)$. This is sometimes called the *trend term*.

Curve Fitting

Fitting a simple function of time such as a polynomial curve (linear, quadratic, etc.), a Gompertz curve, or a logistic curve is a well-known method of dealing with nonseasonal data that contains a trend, particularly yearly data. The global linear trend is the simplest type of polynomial curve. The Gompertz curve can be written in the following format, where α, β, and γ are parameters with $0 < r < 1$:

$$X_t = \alpha \exp\left[\beta \exp\left(-\gamma t\right)\right]$$

This looks quite different but is actually equivalent, provided $t > 0$. The logistic curve is as follows:

$$X_t = a / \left(1 + be^{-ct}\right)$$

Both these curves are S-shaped and approach an asymptotic value as $t \rightarrow \infty$, with the Gompertz curve generally converging slower than the logistic one. Fitting the curves to data may lead to nonlinear simultaneous equations.

For all curves of this nature, the fitted function provides a measure of the trend, and the residuals provide an estimate of local fluctuations where the residuals are the differences between the observations and the corresponding values of the fitted curve.

Removing Trends from a Time Series

Differentiating a given time series until it becomes stationary is a special type of filtering that is particularly useful for removing a trend. You will see that this is an integral part of the Box-Jenkins procedure. For data with a linear trend, a first-order differencing is usually enough to remove the trend.

Mathematically, it looks like this:

$$y(t) = a * t + c$$
$$y(t+1) = a * (t+1) + c$$
$$z(t) = y(t+1) - y(t) = a + c; \text{no trend present in } z(t)$$

A trend can be exponential as well. In this case, you will have to do a logarithmic transformation to convert the trend from exponential to linear.

Mathematically, it looks like this:

$$y(t) = a * \exp(t)$$
$$z(t) = \log(y(t)) = t * \log(a); z(t) \text{ is a linear function of } t$$

Analyzing a Series Containing Seasonality

Many time series, such as airline passenger loads or weather readings, display variations that repeat after a specific time period. For instance, in India, there will always be an increase in airline passenger loads during the holiday of Diwali. This yearly variation is easy to understand and can be estimated if seasonality is of direct interest. Similarly, like trends, if you have a series such as 1, 2, 1, 2, 1, 2, your obvious choices for the next values of the series will be 1 and 2.

The Holt-Winters model is a popular model to realize time series with seasonality and is also known as *exponential smoothing*. The Holt-Winters model has two variations: additive and multiplicative. In the additive model with a single exponential smoothing time series, seasonality is realized as follows:

$$X(t+1) = \alpha * Xt + (1-\alpha)^* St - 1$$

In this model, every point is realized as a weighted average of the previous point and seasonality. So, X(t+1) will be calculated as a function X(t-1) and S(t-2) and square of α. In this way, the more you go on, the α value increases exponentially. This is why it is known as exponential smoothing. The starting value of St is crucial in this method. Commonly, this value starts with a 1 or with an average of the first four observations.

The multiplicative seasonal model time series is as follows:

$$X(t+1) = (b1 + b2 * t)St + noise,$$

Here, b1, often referred to as the *permanent component*, is the initial weight of the seasonality; b2 represents the trend, which is linear in this case.

However, there is no standard implementation of the Holt-Winters model in Python. It is available in R (see Chapter 1 for how R's Holt-Winters model can be called from Python code).

Removing Seasonality from a Time Series

There are two ways of removing seasonality from a time series.

- By filtering
- By differencing

By Filtering

The series $\{x_t\}$ is converted into another called $\{y_t\}$ with the linear operation shown here, where $\{a_r\}$ is a set of weights:

$$Y_t = \sum_{r=-q}^{+s} a_r X_{t+r}$$

To smooth out local fluctuations and estimate the local mean, you should clearly choose the weights so that $\sum a_r = 1$; then the operation is often referred to as a *moving average*. They are often symmetric with s = q and $a_j = a_{-j}$. The simplest example of a symmetric smoothing filter is the simple moving average, for which $a_r = 1 / (2q+1)$ for r = -q, ..., + q.

The smoothed value of x_t is given by the following:

$$Sm(x_t) = \frac{1}{2q+1} \sum_{r=-q}^{+q} X_{t+r}$$

The simple moving average is useful for removing seasonal variations, but it is unable to deal well with trends.

By Differencing

Differencing is widely used and often works well. Seasonal differencing removes seasonal variation.

Mathematically, if time series y(t) contains additive seasonality S(t) with time period T, then:

$$y(t) = a * S(t) = b * t + c$$
$$y(t+T) = aS(t+T) + b * (t+T) + c$$
$$z(t) = y(t+T) - y(t) = b * T + \text{noise term}$$

Similar to trends, you can convert the multiplicative seasonality to additive by log transformation.

Now, finding time period T in a time series is the critical part. It can be done in two ways, either by using an autocorrelation function in the time domain or by using the Fourier transform in the frequency domain. In both cases, you will see a spike in the plot. For autocorrelation, the plot spike will be at lag T, whereas for FT distribution, the spike will be at frequency 1/T.

Transformation

Up to now I have discussed the various kinds of transformation in a time series. The three main reasons for making a transformation are covered in the next sections.

To Stabilize the Variance

The standard way to do this is to take a logarithmic transformation of the series; it brings closer the points in space that are widely scattered.

To Make the Seasonal Effect Additive

If the series has a trend and the volume of the seasonal effect appears to be on the rise with the mean, then it may be advisable to modify the data so as to make the seasonal effect constant from year to year. This seasonal effect is said to be additive. However, if the volume of the seasonal effect is directly proportional to the mean, then the seasonal effect is said to be multiplicative, and a logarithmic transformation is needed to make it additive again.

To Make the Data Distribution Normal

In most probability models, it is assumed that distribution of data is Gaussian or normal. For example, there can be evidence of skewness in a trend that causes "spikes" in the time plot that are all in the same direction.

To transform the data in a normal distribution, the most common transform is to subtract the mean and then divide by the standard deviation. I gave an example of this transformation in the RNN example in Chapter 5; I'll give another in the final example of the current chapter. The logic behind this transformation is it makes the mean 0 and the standard deviation 1, which is a characteristic of a normal distribution. Another popular transformation is to use the logarithm. The major advantage of a logarithm is it reduces the variation and logarithm of Gaussian distribution data that is also Gaussian. Transformation may be problem-specific or domain-specific. For instance, in a time series of an airline's passenger load data, the series can be normalized by dividing by the number of days in the month or by the number of holidays in a month.

Cyclic Variation

In some time series, seasonality is not a constant but a stochastic variable. That is known as *cyclic variation*. In this case, the periodicity first has to be predicted and then has to be removed in the same way as done for seasonal variation.

Irregular Fluctuations

A time series without trends and cyclic variations can be realized as a weekly stationary time series. In the next section, you will examine various probabilistic models to realize weekly time series.

Stationary Time Series

Normally, a time series is said to be stationary if there is no systematic change in mean and variance and if strictly periodic variations have been done away with. In real life, there are no stationary time series. Whatever data you receive by using transformations, you may try to make it somehow nearer to a stationary series.

Stationary Process

A time series is strictly stationary if the joint distribution of $X(t_1),...,X(t_k)$ is the same as the joint distribution of $X(t_1 + \tau),...,X(t_k + \tau)$ for all $t_1,...,t_k,\tau$. If k=1, strict stationary implies that the distribution of $X(t)$ is the same for all t, so provided the first two moments are finite, you have the following:

$$\mu(t) = \mu$$

$$\sigma^2(t) = \sigma^2$$

They are both constants, which do not depend on the value of t.

A weekly stationary time series is a stochastic process where the mean is constant and autocovariance is a function of time lag.

Autocorrelation and the Correlogram

Quantities called *sample autocorrelation coefficients* act as an important guide to the properties of a time series. They evaluate the correlation, if any, between observations at different distances apart and provide valuable descriptive information. You will see that they are also an important tool in model building and often provide valuable clues for a suitable probability model for a given set of data. The quantity lies in the range [-1,1] and measures the forcefulness of the linear association between the two variables. It can be easily shown that the value does not depend on the units in which the two variables are measured; if the variables are independent, then the ideal correlation is zero.

A helpful supplement in interpreting a set of autocorrelation coefficients is a graph called a *correlogram*. The correlogram may be alternatively called the *sample autocorrelation function.*

Suppose a stationary stochastic process X(t) has a mean μ, variance σ^2, auto covariance function (acv.f.) $\gamma(t)$, and auto correlation function (ac.f.) $\rho(\tau)$.

$$\rho(\tau) = \frac{\gamma(\tau)}{\gamma(0)} = \gamma(\tau)/\sigma^2$$

Estimating Autocovariance and Autocorrelation Functions

In the stochastic process, the autocovariance is the covariance of the process with itself at pairs of time points. Autocovariance is calculated as follows:

$$\gamma(h) = \frac{1}{n} \sum_{t=1}^{n-|h|} \left(x_{t+|h|} - \overline{x} \right) \left(x_t - \overline{x} \right), -n < h < n$$

Figure 6-2 shows a sample autocorrelation distribution.

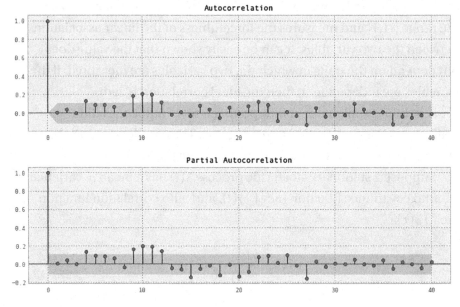

Figure 6-2. *Sample autocorrelations*

Time-Series Analysis with Python

A complement to SciPy for statistical computations including descriptive statistics and estimation of statistical models is provided by Statsmodels, which is a Python package. Besides the early models, linear regression, robust linear models, generalized linear models, and models for discrete data, the latest release of scikits.statsmodels includes some basic tools and models for time-series analysis, such as descriptive statistics, statistical tests, and several linear model classes. The linear model classes include autoregressive (AR), autoregressive moving-average (ARMA), and vector autoregressive models (VAR).

Useful Methods

Let's start with a moving average.

Moving Average Process

Suppose that $\{Z_t\}$ is a purely random process with mean 0 and variance σ_z^2. Then a process $\{X_t\}$ is said to be a moving average process of order q.

$$X_t = \beta_0 Z_t + \beta_1 Z_{t-1} + \ldots + \beta_q Z_{t-q}$$

Here, $\{\beta_i\}$ are constants. The Zs are usually scaled so that $\beta_0 = 1$.

$$E(X_t) = 0$$

$$\mathrm{Var}(X_t) = \sigma_Z^2 \sum_{i=0}^{q} \beta_i^2$$

The Zs are independent.

$$\gamma(k) = \mathrm{Cov}(X_t, X_{t+k})$$
$$= \mathrm{Cov}\left(\beta_0 Z_t + \ldots + \beta_q Z_{t-q}, \beta_0 Z_{t+k} + \ldots + \beta_q Z_{t+k-q}\right)$$
$$= \begin{cases} 0 & k > q \\ \sigma_Z^2 \displaystyle\sum_{i=0}^{q-k} \beta_i \beta_{i+k} & k = 0, 1, \ldots, q \\ \gamma(-k) & k < 0 \end{cases}$$

$$\mathrm{Cov}(Z_s, Z_t) = \begin{cases} \sigma_Z^2 & s = t \\ 0 & s \neq t \end{cases}$$

As $\gamma(k)$ is not dependent on t and the mean is constant, the process is second-order stationary for all values of $\{\beta_i\}$.

$$\rho(k) = \begin{cases} 1 & k = 0 \\ \displaystyle\sum_{i=0}^{q-k} \beta_i \beta_{i+k} \Big/ \displaystyle\sum_{i=0}^{q} \beta_i^2 & k = 1, \ldots, q \\ 0 & k > q \\ \rho(-k) & k < 0 \end{cases}$$

Fitting Moving Average Process

The moving-average (MA) model is a well-known approach for realizing a single-variable weekly stationary time series (see Figure 6-3). The moving-average model specifies that the output variable is linearly dependent on its own previous error terms as well as on a stochastic term. The AR model is called the *moving-average model*, which is a special case and a key component of the ARMA and ARIMA models of time series.

$$X_t = \varepsilon_t + \sum_{i=1}^{p} \varphi_i X_{t-i} + \sum_{i=1}^{q} \theta_i \varepsilon_{t-i} + \sum_{i=1}^{b} \eta_i d_{t-i}$$

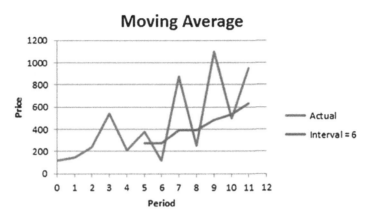

Figure 6-3. *Example of moving average*

Here's the example code for a moving-average model:

```
Please install numpy by following command:
pip install numpy
```

```python
import numpy as np
def moving_average(x, w):
    return np.convolve(x, np.ones(w), 'valid') / w
```

```python
data = np.array([10,5,8,9,15,22,26,11,15,16,18,7])
print(moving_average(data,4))
```

Autoregressive Processes

Suppose $\{Z_t\}$ is a purely random process with mean 0 and variance σ_z^2. After that, a process $\{X_t\}$ is said to be of autoregressive process of order p if you have this:

$$x_t = \alpha_1 x_{t-1} + \ldots \alpha_p x_{t-p} + z_t, or$$

$$x_t = \sum_{i=1}^{p} \alpha_i x_{i-1} + z_t$$

The autocovariance function is given by the following:

$$\gamma(k) = \frac{\alpha^k \sigma_z^2}{(1-\alpha^2)}, k = 0,1,2\ldots, \text{hence}$$

$$\rho_k = \frac{\gamma(k)}{\gamma(0)} = \alpha^k$$

Figure 6-4 shows a time series and its autocorrelation plot of the AR model.

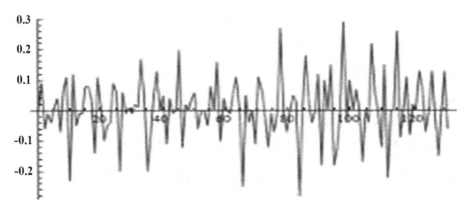

Figure 6-4. *A time series and AR model*

Estimating Parameters of an AR Process

A process is called *weakly stationary* if its mean is constant and the autocovariance function depends only on time lag. There is no weakly stationary process, but it is imposed on time-series data to do some stochastic analysis. Suppose Z(t) is a weak stationary process with mean 0 and constant variance. Then X(t) is an autoregressive process of order p if you have the following:

$$X(t) = a1 \times X(t-1) + a2 \times X(t-2) + \ldots + ap \times X(t-p) + Z(t), \text{where } a \in R \text{ and } p \in II$$

Now, E[X(t)] is the expected value of X(t).

$$\text{Covariance}(X(t),X(t+h)) = E\left[(X(t)\text{-}E[X(t)]) * (X(t+h)\text{-}E[X(t+h)])\right]$$
$$= E\left[(X(t)\text{-}m) * (X(t+h)\text{-}m)\right]$$

If X(t) is a weak stationary process, then:

$$E[X(t)] = E[X(t+h)] = m(\text{constant})$$
$$= E[X(t) * X(t+h)] - m^2 = c(h)$$

Here, m is constant, and cov[X(t),X(t+h)] is the function of only h(c(h)) for the weakly stationary process. c(h) is known as autocovariance.

Similarly, the correlation $(X(t),X(t+h) = \rho(h) = r(h) = c(h) = c(0)$ is known as *autocorrelation*.

If X(t) is a stationary process that is realized as an autoregressive model, then:

$$X(t) = a1 * X(t-1) + a2 * X(t-2) + \ldots + ap * X(t-p) + Z(t)$$

Correlation(X(t),X(t)) = a1 * correlation (X(t),X(t-1)) + ... + ap * correlation (X(t),X(t-p))+0

As covariance, (X(t),X(t+h)) is dependent only on h, so:

$$r0 = a1 * r1 + a2 * r2 + ... + ap * rp$$

$$r1 = a1 * r0 + a2 * r1 + ... + ap * r(p\text{-}1)$$

So, for an n-order model, you can easily generate the n equation and from there find the n coefficient by solving the n equation system.

In this case, realize the data sets only in the first-order and second-order autoregressive model and choose the model whose mean of residual is less. For that, the reduced formulae are as follows:

- *First order:* a1 = r1
- *Second order:* $a1 = r_1\left(1 - r_2\right) \div \left(1 - r_1^2\right), a2 = \left(r_2 - r_1^2\right) \div \left(1 - r_1^2\right)$

Here is some example code for an autoregressive model:

```
Please download the required CSC file from https://github.com/
pratips/book-Advanced-Data-Analytics-Using-Python and install
the Python libraries using the pip command.
```

```python
from pandas import read_csv
from matplotlib import pyplot
from statsmodels.tsa.ar_model import AutoReg
from sklearn.metrics import mean_squared_error
from math import sqrt

# load dataset
series = read_csv('temperatures.csv', header=0, index_col=0,
parse_dates=True, squeeze=True)
```

```
# split dataset
X = series.values
train, test = X[1:len(X)-7], X[len(X)-7:]

# train autoregression
model = AutoReg(train, lags=29)
model_fit = model.fit()
print('Coefficients: %s' % model_fit.params)

# make predictions
predictions = model_fit.predict(start=len(train),
end=len(train)+len(test)-1, dynamic=False)
for i in range(len(predictions)):
print('predicted=%f, expected=%f' % (predictions[i], test[i]))
rmse = sqrt(mean_squared_error(test, predictions))
print('Test RMSE: %.3f' % rmse)

# plot results
pyplot.plot(test)
pyplot.plot(predictions, color='red')
pyplot.show()
```

Mixed ARMA Models

Mixed ARMA models are a combination of MA and AR processes. A mixed autoregressive/moving average process containing p AR terms and q MA terms is said to be an ARMA process of order (p,q). It is given by the following:

$$X_t = \alpha_1 X_{t-1} + \ldots + \alpha_p X_{t-p} + Z_t + \beta_1 Z_{t-1} + \ldots \beta_q Z_{t-q}$$

The following example code was taken from the stat model site to realize time-series data as an ARMA model.

Please install matplotlib, statsmodels, and numpy using the pip command before running the code.

```
import matplotlib.pyplot as plt
import numpy as np
from statsmodels.tsa.arima_process import ArmaProcess
import statsmodels.api as sm

np.random.seed(1234)
data = np.array([1, .85, -.43, -.63, .8])
parameter = np.array([1, .41])
model = ArmaProcess(data, parameter)
fig = plt.figure(figsize=(12, 8))
ax = fig.add_subplot(111)
ax.plot(model.generate_sample(nsample=50))
plt.show()
```

Here is how to estimate parameters of an ARMA model:

1. After specifying the order of a stationary ARMA process, you need to estimate the parameters.

2. Assume the following:

    - The model order (p and q) is known.

    - The data has zero mean.

3. If step 2 is not a reasonable assumption, you can subtract the sample mean Y and fit a 0 mean ARMA model, as in $\emptyset(B)X_t = \theta(B)a_t$ where $X_t = Y_t - Y$. Then use $X_t + Y$ as the model for Y_t.

Integrated ARMA Models

To fit a stationary model such as the one discussed earlier, it is imperative to remove nonstationary sources of variation. Differencing is widely used for econometric data. If X_t is replaced by $\nabla^d X_t$, then you have a model capable of describing certain types of nonstationary series.

$$Y_t = (1 - L)^d X_t$$

These are the estimating parameters of an ARIMA model:

- ARIMA models are designated by the level of autoregression, integration, and moving averages.

- This does not assume any pattern uses an iterative approach of identifying a model.

- The model "fits" if residuals are generally small, randomly distributed, and, in general, contain no useful information.

Here is the example code for an ARIMA model.

Please download the required CSV file from https://github.com/pratips/book-Advanced-Data-Analytics-Using-Python and install the Python libraries using the pip command.

```
# importing libraries

from pandas import read_csv
from matplotlib import pyplot
from statsmodels.tsa.arima.model import ARIMA
from sklearn.metrics import mean_squared_error

# readfing csv
```

```
series = read_csv('temperatures.csv', header=0, index_col=0,
parse_dates=True, squeeze=True)
P = series.values
size = int(len(P) * 0.9)

#splitting into train test set

train, test = P[0:size], P[size:len(P)]
history = [p for p in train]
predictions = list()
for t in range(len(test)):

    # fitting model

    model = ARIMA(history, order=(5,1,0))
    model_fit = model.fit()
    output = model_fit.forecast()
    yhat = output[0]
    predictions.append(yhat)
    obs = test[t]
    history.append(obs)
    print('predicted=%f, expected=%f' % (yhat, obs))
error = mean_squared_error(test, predictions)
print('Test MSE: %.3f' % error)

# plotting results

pyplot.plot(test)
pyplot.plot(predictions, color='red')
pyplot.show()
```

The Fourier Transform

The representation of nonperiodic signals by everlasting exponential signals can be accomplished by a simple limiting process, and I will illustrate that nonperiodic signals can be expressed as a continuous sum (integral) of everlasting exponential signals. Say you want to represent the nonperiodic signal g(t). Realizing any nonperiodic signal as a periodic signal with an infinite time period, you get the following:

$$g(t) = \frac{1}{2\pi} \int_{-\infty}^{\infty} G(\omega) e^{j\omega t} d\omega$$

$$G(n\Delta\omega) = \lim_{T \to \infty} \int_{-T_0/2}^{T_0/2} g_p(t) e^{-jn\Delta\omega t} dt$$

Hence:

$$G(\omega) = \int_{-\infty}^{\infty} g(t) e^{-j\omega t} dt$$

G(w) is known as a Fourier transform of g(t).

Here is the relation between autocovariance and the Fourier transform:

$$\gamma(0) = \sigma_X^2 = \int_0^{\pi} dF(\omega) = F(\pi)$$

An Exceptional Scenario

In the airline or hotel domain, the passenger load of month t is less correlated with data of t-1 or t-2 month, but it is more correlated for t-12 month. For example, the passenger load in the month of Diwali (October) is more correlated with last year's Diwali data than with the same year's August and September data. Historically, the pick-up model is used to predict this kind of data. The pick-up model has two variations.

In the additive pick-up model,

$$X(t) = X(t-1) + \left[X(t-12) - X(t-13) \right]$$

In the multiplicative pick-up model,

$$X(t) = X(t-1) * \left[X(t-12) / X(t-13) \right]$$

Studies have shown that for this kind of data the neural network–based predictor gives more accuracy than the time-series model.

In high-frequency trading in investment banking, time-series models are too time-consuming to capture the latest pattern of the instrument. So, they on the fly calculate dX/dt and d2X/dt2, where X is the price of the instruments. If both are positive, they blindly send an order to buy the instrument. If both are negative, they blindly sell the instrument if they have it in their portfolio. But if they have an opposite sign, then they do a more detailed analysis using the time-series data.

As I stated earlier, there are many scenarios in time-series analysis where R is a better choice than Python. So, here is an example of time-series forecasting using R. The beauty of the auto.arima model is that it automatically finds the order, trends, and seasonality of the data and fits the model. In the forecast, we are printing only the mean value, but the model provides the upper limit and the lower limit of the prediction in forecasting.

Please install the pmdarima library by using the following command:

```
pip install pmdarima
```

Once you install, you can try running the following code based on pmdarima's manual. It loads a data set available in the package and then tries to fit auto arima and predicts as well.

```
import numpy as np
import pmdarima as pm
from pmdarima.datasets import load_wineind

# loading dataset
wineind = load_wineind().astype(np.float64)
# fitting a stepwise model:
stepwise_fit = pm.auto_arima(wineind, start_p=1, start_q=1,
max_p=3, max_q=3, m=12, start_P=0, seasonal=True, d=1, D=1,
trace=True, error_action='ignore', suppress_warnings=True,
stepwise=True)

stepwise_fit.summary()
predicted_15 = stepwise_fit.predict(n_periods=15)
print(predicted_15)
```

Missing Data

One important aspect of time series and many other data analysis work is figuring out how to deal with missing data. In the previous code, you fill in the missing record with the average value. This is fine when the number of missing data instances is not very high. But if it is high, then the average of the highest and lowest values is a better alternative.

Summary

Following feature engineering, we'll go over some basic statistics, particularly time-series models. One thing to keep in mind is that you may apply any supervised machine learning model on time-series data if you transform feature vectors (1xn) to the matrix (kxn), where each k element of each row is the latest k observation of the first column.

CHAPTER 7

Analytics at Scale

In recent decades, a revolutionary change has taken place in the field of analytics technology because of big data. Data is being collected from a variety of sources, so technology has been developed to analyze this data in a distributed environment, even in real time.

Hadoop

The revolution started with the development of the Hadoop framework, which has two major components, namely, MapReduce programming and the HDFS file system.

MapReduce Programming

MapReduce is a programming style inspired by functional programming to deal with large amounts of data. The programmer can process big data using MapReduce code without knowing the internals of the distributed environment. Before MapReduce, frameworks like Condor did parallel computing on distributed data. But the main advantage of MapReduce is that it is RPC based. The data does not move; on the contrary, the code jumps to different machines to process the data. In the case of big data, it is a huge savings of network bandwidth as well as computational time.

A MapReduce program has two major components: the mapper and the reducer. In the mapper, the input is split into small units. Generally, each line of input file becomes an input for each map job. The mapper

© Sayan Mukhopadhyay, Pratip Samanta 2023
S. Mukhopadhyay and P. Samanta, *Advanced Data Analytics Using Python*,
https://doi.org/10.1007/978-1-4842-8005-8_7

processes the input and emits a key-value pair to the reducer. The reducer receives all the values for a particular key as input and processes the data for final output.

The following pseudocode is an example of counting the frequency of words in a document:

```
map(String key, String value):
// key: document name
// value: document contents
for each word w in value:
EmitIntermediate(w, "1");
reduce(String key, Iterator values):
 // key: a word
 // values: a list of counts
int result = 0;
for each v in values:
result += ParseInt(v);
Emit(AsString(result));
```

Partitioning Function

Sometimes it is required to send a particular data set to a particular reduce job. The partitioning function solves this purpose. For example, in the previous MapReduce example, say the user wants the output to be stored in sorted order. Then he mentions the number of the reduce job 32 for 32 letters, and in the practitioner he returns 1 for the key starting with a, 2 for b, and so on. Then all the words that start with the same letters go to the same reduce job. The output will be stored in the same output file, and because MapReduce assures that the intermediate key-value pairs are processed in increasing key order, within a given partition, the output will be stored in sorted order.

Combiner Function

The combiner is a facility in MapReduce where partial aggregation is done in the map phase. Not only does it increase the performance, but sometimes it is essential to use if the data set so huge that the reducer is throwing a stack overflow exception. Usually the reducer and combiner logic are the same, but this might be necessary depending on how MapReduce deals with the output.

To implement this word count example, we will follow a particular design pattern. There will be a root RootBDAS (BDAS stands for Big Data Analytic System) class that has two abstract methods: a mapper task and a reducer task. All child classes implement these mapper and reducer tasks. The main class will create an instance of the child class using reflection, and in MapReduce map functions call the mapper task of the instance and the reducer function of the reducer task. The major advantages of this pattern are that you can do unit testing of the MapReduce functionality and that it is adaptive. Any new child class addition does not require any changes in the main class or unit testing. You just have to change the configuration. Some code may need to implement combiner or partitioner logics. They have to inherit the ICombiner or IPartitioner interface.

Figure 7-1 shows a class diagram of the system.

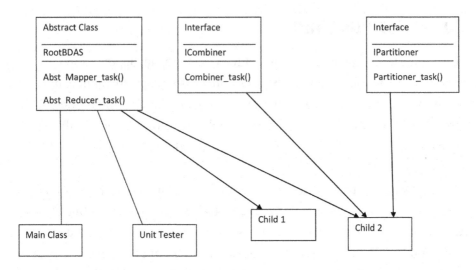

Figure 7-1. *The class diagram*

Here is the RootBDAS class:

```
import java.util.ArrayList;
import java.util.HashMap;
/**
 *
 */
/**
 * @author SayanM
 *
 */
public abstract class RootBDAS {
        abstract  HashMap<String, ArrayList<String>>  mapper_
        task(String line);
        abstract  HashMap<String, ArrayList<String>>  reducer_
        task(String key, ArrayList<String> values);
}
```

Here is the child class:

```java
import java.util.ArrayList;
import java.util.HashMap;
/**
 *
 */
/**
 * @author SayanM
 *
 */
public final class WordCounterBDAS extends RootBDAS{
        @Override
        HashMap<String, ArrayList<String>> mapper_
        task(String line) {
                // TODO Auto-generated method stub
                String[] words = line.split(" ");
                HashMap<String, ArrayList<String>> result = new
                HashMap<String, ArrayList<String>>();
                for(String w : words)
                {
                        if(result.containsKey(w))
                        {
                                ArrayList<String> vals =
                                result.get(w);
                                vals.add("1");
                                result.put(w, vals);
                        }
                        else
                        {
                                ArrayList<String> vals = new
                                ArrayList<String>();
```

```java
                            vals.add("1");
                            result.put(w, vals);
                    }
            }
            return result;

        }
        @Override
        HashMap<String, ArrayList<String>> reducer_task(String
        key, ArrayList<String> values) {
                // TODO Auto-generated method stub
                HashMap<String, ArrayList<String>> result = new
                HashMap<String, ArrayList<String>>();
                ArrayList<String> tempres = new
                ArrayList<String>();
                tempres.add(values.size()+ "");
                result.put(key, tempres);
                return result;

        }
}
```

Here is the WordCounterBDAS utility class:

```java
import java.util.ArrayList;
import java.util.HashMap;
/**
 *
 */
/**
 * @author SayanM
 *
 */
public final class WordCounterBDAS extends RootBDAS{
```

```java
@Override
HashMap<String, ArrayList<String>> mapper_
task(String line) {
        // TODO Auto-generated method stub
        String[] words = line.split(" ");
        HashMap<String, ArrayList<String>> result = new
        HashMap<String, ArrayList<String>>();
        for(String w : words)
        {
            if(result.containsKey(w))
            {
                ArrayList<String> vals = result.get(w);
                 vals.add("1");
                 result.put(w, vals);
            }
            else
            {
                ArrayList<String> vals = new
                ArrayList<String>();
                vals.add("1");
                result.put(w, vals);
            }
        }
        return result;
}
@Override
HashMap<String, ArrayList<String>> reducer_task(String
key, ArrayList<String> values) {
        // TODO Auto-generated method stub
HashMap<String, ArrayList<String>> result = new
HashMap<String, ArrayList<String>>();
        ArrayList<String> tempres = new ArrayList<String>();
```

```
                tempres.add(values.size()+ "");
                result.put(key, tempres);
                return result;
        }
}
```

Here is the MainBDAS class:

```
import java.io.IOException;
import java.util.ArrayList;
import java.util.HashMap;
import org.apache.hadoop.conf.Configuration;
import org.apache.hadoop.fs.Path;
import org.apache.hadoop.io.LongWritable;
import org.apache.hadoop.io.Text;
import org.apache.hadoop.mapreduce.Job;
import org.apache.hadoop.mapreduce.Mapper;
import org.apache.hadoop.mapreduce.Reducer;
import org.apache.hadoop.mapreduce.lib.input.FileInputFormat;
import org.apache.hadoop.mapreduce.lib.input.TextInputFormat;
import org.apache.hadoop.mapreduce.lib.output.FileOutputFormat;
import org.apache.hadoop.mapreduce.lib.output.TextOutputFormat;
/**
 *
 */
/**
 * @author SayanM
 *
 */
public class MainBDAS {
        public static class MapperBDAS extends
        Mapper<LongWritable, Text, Text, Text> {
```

```java
    protected void map(LongWritable key, Text value,
Context context)
            throws IOException,
            InterruptedException {
        String classname = context.
        getConfiguration().get("classname");
        try {
            RootBDAS instance = (RootBDAS)
            Class.forName(classname).
            getConstructor().newInstance();
            String line = value.toString();
            HashMap<String, ArrayList<String>>
            result = instance.mapper_task(line);
            for(String k : result.keySet())
            {
                    for(String v : result.get(k))
                    {
                            context.write(new
                            Text(k), new Text(v));
                    }
            }
        } catch (Exception e) {
            // TODO Auto-generated catch block
            e.printStackTrace();
        }
    }
}
public static class ReducerBDAS extendsReducer<Text,
Text, Text, Text> {
    protected void reduce(Text key,
    Iterable<Text> values,
```

```
                Context context) throws IOException,
                InterruptedException {
        String classname = context.
        getConfiguration().get("classname");
        try {
                RootBDAS instance = (RootBDAS)
                Class.forName(classname).
                getConstructor().newInstance();
                ArrayList<String> vals = new
                ArrayList<String>();
                for(Text v : values)
                {
                        vals.add(v.toString());
                }
                HashMap<String, ArrayList<String>>
                result = instance.reducer_task(key.
                toString(), vals);
                for(String k : result.keySet())
                {
                        for(String v : result.get(k))
                        {
                                context.write(new
                                Text(k), new Text(v));
                        }
                }
        } catch (Exception e) {
                // TODO Auto-generated catch block
                e.printStackTrace();
        }
    }
}
```

```
public static void main(String[] args) throws Exception {
    // TODO Auto-generated method stub
  String classname = Utility.getClassName(Utility.
  configpath);
    Configuration con = new Configuration();
    con.set("classname", classname);
    Job job = new Job(con);
    job.setJarByClass(MainBDAS.class);
    job.setJobName("MapReduceBDAS");
    job.setOutputKeyClass(Text.class);
    job.setOutputValueClass(Text.class);
    job.setInputFormatClass(TextInputFormat.class);
    job.setOutputFormatClass(TextOutputFormat.class);
    FileInputFormat.setInputPaths(job, new
    Path(args[0]));
    FileOutputFormat.setOutputPath(job, new
    Path(args[1]));
    job.setMapperClass(MapperBDAS.class);
    job.setReducerClass(ReducerBDAS.class);
    System.out.println(job.waitForCompletion(true));
  }
}
```

To test the example, you can use this unit testing class:

```
import static org.junit.Assert.*;
import java.util.ArrayList;
import java.util.HashMap;
import org.junit.Test;
public class testBDAS {
    @Test
    public void testMapper() throws Exception{
        String classname = Utility.getClassName(Utility.
        testconfigpath);
```

```
            RootBDAS instance = (RootBDAS) Class.
            forName(classname).getConstructor().
            newInstance();
            String line = Utility.getMapperInput(Utility.
            testconfigpath);
            HashMap<String, ArrayList<String>> actualresult =
            instance.mapper_task(line);
            HashMap<String, ArrayList<String>> expectedresult =
            Utility.getMapOutput(Utility.testconfigpath);
            for(String key : actualresult.keySet())
            {
                    boolean haskey = expectedresult.
                    containsKey(key);
                    assertEquals(true, haskey);
                    ArrayList<String> actvals = actualresult.
                    get(key);
                    for(String v : actvals)
                    {
                            boolean hasval = expectedresult.
                            get(key).contains(v);
                            assertEquals(true, hasval);
                    }
            }
    }
    @Test
    public void testReducer(){
            fail();
    }
}
```

Finally, here are the interfaces:

```java
import java.util.ArrayList;
import java.util.HashMap;
public interface ICombiner {
        HashMap<String, ArrayList<String>>  combiner_task(String
        key, ArrayList<String> values);
}
public interface IPartitioner {
        public int  partitioner_task(String line);
}
```

HDFS File System

Other than MapReduce, HDFS is the second component in the
Hadoop framework. It is designed to deal with big data in a distributed
environment for general-purpose low-cost hardware. HDFS is built on top
of the Unix POSSIX file system with some modifications, with the goal of
dealing with streaming data.

The Hadoop cluster consists of two types of host: the name node and
the data node. The name node stores the metadata, controls execution,
and acts like the master of the cluster. The data node does the actual
execution; it acts like a slave and performs instructions sent by the
name node.

MapReduce Design Pattern

MapReduce is an archetype for processing the data that resides in
hundreds of computers. There are some design patterns that are common
in MapReduce programming.

Summarization Pattern

In summary, the reducer creates the summary for each key (see Figure 7-2). The practitioner can be used if you want to sort the data or for any other purpose. The word count is an example of the summarizer pattern. This pattern can be used to find the minimum, maximum, and count of data or to find the average, median, and standard deviation.

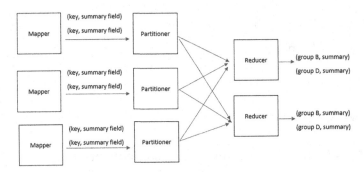

Figure 7-2. *Details of the summarization pattern*

Filtering Pattern

In MapReduce, filtering is done in a divide-and-conquer way (Figure 7-3). Each mapper job filters a subset of data, and the reducer aggregates the filtered subset and produces the final output. Generating the top N records, searching data, and sampling data are the common use cases of the filtering pattern.

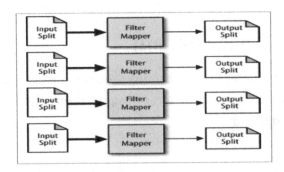

Figure 7-3. *Details of the filtering pattern*

198

Join Patterns

In MapReduce, joining (Figure 7-4) can be done on the map side or the reduce side. For the map side, the join data sets that will be joined should exist in the same cluster; otherwise, the reduce-side join is required. The join can be an outer join, inner join, or anti-join.

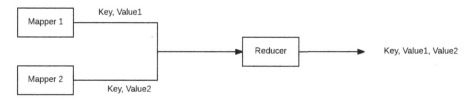

Figure 7-4. *Details of the join pattern*

The following code is an example of the reducer-side join:

```
package MapreduceJoin;
import java.io.IOException;
import java.util.ArrayList;
import java.util.Iterator;
import org.apache.hadoop.fs.Path;
import org.apache.hadoop.io.LongWritable;
import org.apache.hadoop.io.Text;
import org.apache.hadoop.mapred.JobConf;
import org.apache.hadoop.mapred.MapReduceBase;
import org.apache.hadoop.mapred.OutputCollector;
import org.apache.hadoop.mapred.Reporter;
import org.apache.hadoop.mapred.lib.MultipleInputs;
import org.apache.hadoop.mapreduce.Job;
import org.apache.hadoop.mapred.Mapper;
import org.apache.hadoop.mapred.Reducer;
import org.apache.hadoop.mapreduce.lib.output.FileOutputFormat;
import org.apache.hadoop.mapred.TextInputFormat;
```

```
@SuppressWarnings("deprecation")
public class MapreduceJoin {
//////////////////////////////////////////////////////
        @SuppressWarnings("deprecation")
        public static class JoinReducer extends MapReduceBase
        implements Reducer<Text, Text, Text, Text>
        {
                public void reduce(Text key, Iterator<Text>
                values, OutputCollector<Text, Text> output,
                Reporter reporter) throws IOException
                {
                        ArrayList<String> translist = new
                        ArrayList<String>();
                        String secondvalue = "";
                        while (values.hasNext())
                        {
                                String currValue = values.next().
                                toString().trim();
                                if(currValue.contains("trans:")){
                                        String[] temp = currValue.
                                        split("trans:");
                                        if(temp.length > 1)
                                                translist.add(temp[1]);
                                }
                                if(currValue.contains("sec:"))
                                {
                                        String[] temp = currValue.
                                        split("sec:");
                                        if(temp.length > 1)
                                                secondvalue = temp[1];
                                }
                        }
```

```
                for(String trans : translist)
                {
                        output.collect(key, new Text(trans
                        +'\t' + secondvalue));
                }
        }
}
///////////////////////////////////////////////////////
@SuppressWarnings("deprecation")
public static class TransactionMapper extends MapReduceBase
implements Mapper<LongWritable, Text, Text, Text>
{
        int index1 = 0;
        public void configure(JobConf job) {
                index1 = Integer.parseInt(job.
                get("index1"));
        }
        public void map(LongWritable key, Text value,
        OutputCollector<Text, Text> output, Reporter
        reporter) throws IOException
         {
                String line = value.toString().trim();
                if(line=="") return;
          String splitarray[] = line.split("\t");
          String id = splitarray[index1].trim();
          String ids = "trans:" + line;
          output.collect(new Text(id), new Text(ids));
        }
}
```

```
/////////////////////////////////////////////////////////////
@SuppressWarnings("deprecation")
public static class SecondaryMapper extends MapReduceBase
implements Mapper<LongWritable, Text, Text, Text>
{
        int index2 = 0;
        public void configure(JobConf job) {
                index2 = Integer.parseInt(job.get("index2"));
        }
        public void map(LongWritable key, Text value,
        OutputCollector<Text, Text> output, Reporter
        reporter) throws IOException
        {
                String line = value.toString().trim();
                if(line=="") return;
          String splitarray[] = line.split("\t");
          String id = splitarray[index2].trim();
          String ids = "sec:" + line;
          output.collect(new Text(id), new Text(ids));
        }
}
/////////////////////////////////////////////////////////////
    @SuppressWarnings({ "deprecation", "rawtypes",
    "unchecked" })
    public static void main(String[] args)
    throws IOException, ClassNotFoundException,
    InterruptedException {
            // TODO Auto-generated method stub
            JobConf conf = new JobConf();
            conf.set("index1", args[3]);
            conf.set("index2", args[4]);
            conf.setReducerClass(JoinReducer.class);
```

```
        MultipleInputs.addInputPath(conf, new
        Path(args[0]), TextInputFormat.class, (Class<?
        extends org.apache.hadoop.mapred.Mapper>)
        TransactionMapper.class);
        MultipleInputs.addInputPath(conf, new
        Path(args[1]), TextInputFormat.class, (Class<?
        extends org.apache.hadoop.mapred.Mapper>)
        SecondaryMapper.class);
        Job job = new Job(conf);
        job.setJarByClass(MapreduceJoin.class);
        job.setJobName("MapReduceJoin");
        job.setOutputKeyClass(Text.class);
        job.setOutputValueClass(Text.class);
        FileOutputFormat.setOutputPath(job, new
        Path(args[2]));
        System.out.println(job.waitForCompletion(true));
    }
}
```

A Notes on Functional Programming

MapReduce is functional programming. Functional programming is a type of programming in which all variables are immutable, making parallelization easy and avoiding race conditions. Normal Python allows functional programming.

Here is an example:

```
employees = [{
    'name': 'Jane',
    'salary': 90000,
    'job_title': 'developer'
}, {
    'name': 'Bill',
```

```
    'salary': 50000,
    'job_title': 'writer'
}, {
    'name': 'Kathy',
    'salary': 120000,
    'job_title': 'executive'
}, {
    'name': 'Anna',
    'salary': 100000,
    'job_title': 'developer'
}, {
    'name': 'Dennis',
    'salary': 95000,
    'job_title': 'developer'
}, {
    'name': 'Albert',
    'salary': 70000,
    'job_title': 'marketing specialist'
}]

#find the average salary of develper

developers = mean(map( lambda e: e.salary if e.job_title
=='developer'))
```

Spark

After Hadoop, Spark is the next and latest revolution in big data technology. The major advantage of Spark is that it gives a unified interface to the entire big data stack. Previously, if you needed a SQL–like interface for big data, you would use Hive. If you needed real-time data processing, you would use Storm. If you wanted to build a machine learning model, you would use Mahout. Spark brings all these facilities under one umbrella. In addition, it enables in-memory computation of big data, which makes the processing very fast. Figure 7-5 describes all the components of Spark.

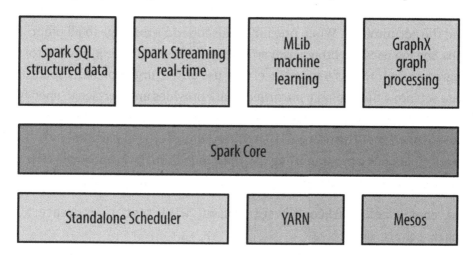

Figure 7-5. *The components of Spark*

Spark Core is the fundamental component of Spark. It can run on top of Hadoop or stand-alone. It abstracts the data set as a resilient distributed data set (RDD). RDD is a collection of read-only objects. Because it is read-only, there will not be any synchronization problems when it is shared with multiple parallel operations. Operations on RDD are lazy. There are two types of operations happening on RDD: transformation and action. In transformation, there is no execution happening on a data set. Spark only stores the sequence of operations as a directed acyclic graph called a *lineage*. When an action is called, then the actual execution takes place. After the first execution, the result is cached in memory. So, when a new execution is called, Spark makes a traversal of the lineage graph and makes maximum reuse of the previous computation, and the computation for the new operation becomes the minimum. This makes data processing very fast and also makes the data fault tolerant. If any node fails, Spark looks at the lineage graph for the data in that node and easily reproduces it.

One limitation of the Hadoop framework is that it does not have any message-passing interface in parallel computation. But there are several use cases where parallel jobs need to talk with each other. Spark achieves this using two kinds of shared variable. They are the broadcast variable and the accumulator. When one job needs to send a message to all other jobs, the job uses the broadcast variable, and when multiple jobs want to aggregate their results to one place, they use an accumulator. RDD splits its data set into a unit called a *partition*. Spark provides an interface to specify the partition of the data, which is very effective for future operations like join or find. The user can specify the storage type of partition in Spark. Spark has a programming interface in Python, Java, and Scala. The following code is an example of a word count program in Spark:

```
val conf = new SparkConf().setAppName("wiki_test") // create a
spark config object
val sc = new SparkContext(conf) // Create a spark context
val data = sc.textFile("/path/to/somedir") // Read files from
"somedir" into an RDD of (filename, content) pairs.
val tokens = data.flatMap(_.split(" ")) // Split each file into
a list of tokens (words).
val wordFreq = tokens.map((_, 1)).reduceByKey(_ + _) // Add a
count of one to each token, then sum the counts per word type.
wordFreq.sortBy(s => -s._2).map(x => (x._2, x._1)).top(10)
// Get the top 10 words. Swap word and count to sort by count.
```

On top of Spark Core, Spark provides the following:

- Spark SQL, which is a SQL interface through the command line or a database connector interface. It also provides a SQL interface for the Spark data frame object.

- Spark Streaming, which enables you to process streaming data in real time.

- MLib, a machine learning library to build analytical models on Spark data.

- GraphX, a distributed graph processing framework.

PySpark

PySpark in a Python interface for Apache Spark. Using simple Python APIs, we can write a Spark application. All the major features of Spark, namely, Spark SQL, MLib, and the data frame, are supported by PySpark. The following are the installation steps:

1. First, we need Java Runtime Environment (JRE) in order to run Spark. It is recommended to install JRE from following link: `https://adoptium.net/temurin/releases/?version=8`. Also, you can install JRE from other distribution.
   The installation will automatically create `JAVA_HOME`. If it does not, we need to set `JAVA_HOME` as the system variable; the value should be the location of the Java installation folder such as `JAVA_HOME=C:\Program Files\Eclipse Adoptium\jdk-8.0.345.1-hotspot\`.

2. Then we need to install PySpark using the command `pip install pyspark`.
   For this we need to have Python 3.7 and above installed. After that, we need to set env `PYSPARK_PYTHON` as the location of Python. Here's an example: `C:\Users\...\Programs\Python\Python39\python.exe`.

3. Also, install PyArrow by using `pip install pyarrow` to use the Pandas API in PySpark.

Now, we are set to write the code in PySpark. The CSV file is the same as the one we used earlier.

```python
from pyspark.sql import SparkSession
from datetime import datetime, date
import pandas as pd
from pyspark.sql import Row
import pyspark.pandas as ps
import numpy as np

# this is create Spark session
spark = SparkSession.builder.getOrCreate()
print(spark)

# Create series
s = ps.Series([11, np.nan, 51, 16, 82])
print(s)

# Create data frame
pandas_df = pd.DataFrame({
    'C1': [11, 22, 31],
    'C2': [12., 13., 14.],
    'C3': ['text1', 'text2', 'text3'],
    'C4': [date(2010, 1, 1), date(2011, 2, 1), date
(2012, 3, 1)]
})
df = spark.createDataFrame(pandas_df)
df.show()
df.printSchema()

#Read csv in using Spark
df = spark.read.csv ('book\\ch4\\CC_GENERAL.csv', header=True)
df.show()
psdf = df.pandas_api()
print(psdf.describe())
```

Updatable Machine Learning and Spark Memory Model

In this section, we discuss a new machine learning topic: how to develop an updatable machine learning model, which is a basic requirement for a model to become scalable.

When we train a model in machine learning, we data train the model from scratch, which implies that all the information from prior training iterations is swapped out. For example, if we train a model with data from the first week of the month, it will predict based on the first week's data, and if we train the same model object with data from the second week, it will predict based on the second week's data. It begins to forecast based only on data from the second week of the month. In the model, there is no trace of first-week data. However, in the real world, there are many scenarios when we require the model to forecast based on first- and second-week data after training, which implies that each training iteration will update the model rather than erase prior training information.

This type of model is possible when the model is a function of X and for any two data sets N1 and N2.

$$X(N1) + X(N2) = X(N1+N2)$$

Examples of this type of function are count, max, min, and sum. Any probability-based model may be divided into sum and count and therefore converted into an updatable model. The Bayesian classifier was used as a model in our application, and the Spark code in the next is our own implementation of the classifier.

The algorithm's goal is to determine the conditional probability that a Node(video or content) will be observed for a specific feature value.

Assume we're calculating the score for node 10033207 with the feature identifier (device ID) and the feature value 49 (choose any device ID according to your choice).

- Feature count = Number of records where that device ID (49) is present

- Conditional count = Number of records with `is_completed` equal to 1, `identifier` equal to 49, and `node` equal to 10033207

The ratio of the conditional count and feature count will represent the actual probability, but we are not dividing this stage because the model will not be updatable.

In each execution of the following code, the following happens:

1. Create a user-defined function to convert the IP address to geographic locations.

2. Read the name of the target channel from the config file.

3. In a data frame, load the lines from the application server log except for the line that contains the newlog.

4. Filter the logline that includes the channel name.

5. Read important column names from the config and selecting them from the data frame.

6. Connect to MySQL and reading the threshold timestamp from the `threshold` table.

7. In MySQL, filter the data frame column with a timestamp more than the threshold, get the max timestamp from the selected logline, and set that as the new threshold for the next iteration.

8. The content or node IDs that are clicked are then saved in the database.

9. Store the popular contents or node in the database.

10. Each line contains the user's next watch node, which is obtained by self-joining the user's previous watch node.

11. Then calculate the ratio of watch time and duration of video as a score.

12. If the score is greater than the channel's threshold defined in config, mark it 1; otherwise, mark it 0.

13. Then, as stated in Chapter 3, calculate the conditional probability of a user watching a video that is longer than the threshold in the form of a count.

```
from pyspark.sql import SQLContext
from pyspark import SparkConf, SparkContext
from pyspark.sql.functions import col
import pyspark.sql.functions as F
from pyspark.sql.functions import when
from pyspark.sql.functions import lit
from pyspark.sql.functions import trim
from pyspark.sql.functions import udf
from pyspark.sql.types import *
import pyspark.sql
import GeoIP

def ip_to_geo(ip):
    gi = GeoIP.open("/home/hadoop/GeoLiteCity.dat", GeoIP.
    GEOIP_STANDARD)
    gir = gi.record_by_name(ip)
    del gi
    if gir is not None:
        if gir['city'] is None:
```

```python
            gir['city']=' '
        if gir['region_name'] is None:
            gir['region_name']=' '
        if gir['country_name'] is None:
            gir['country_name']=' '
        return gir['city'] + ',' + gir['region_name'] + ','
        + gir['country_name']
    else:
        return " , , "

geo_udf = udf(ip_to_geo,StringType())

sc = SparkContext(appName="BuildModelContent")
sqlContext = SQLContext(sc)

channel_config = sqlContext.read.json('/tmp/config_
channels.json')
channels_list = channel_config.select("channels").collect()
channels = [(row.channels) for row in channels_list]

lines = sc.textFile("/tmp/log2.txt")
rows = lines.filter(lambda line: 'newlog' not in line)

config = sqlContext.read.json('/tmp/config.json',
multiLine=True)
lim = config.select('limit').collect()[0]['limit']

rows = rows.filter(lambda row: any(c in row for c in channels))
rdd = rows.map(lambda x: x.split('|'))
#r_f = rdd.first()
df_log = rdd.toDF()      #filter(lambda row: len(row) ==
len(r_f)).toDF()
rdd.unpersist()
sqlContext.clearCache()
df_log.show()
```

```
parameter_config = sqlContext.read.json('/tmp/config_
parameters.json')
parameter_config.show()
column_index_list = parameter_config.select("Index").collect()
column_name_list = parameter_config.select("Name").collect()
column_index =  [col(row.Index) for row in column_index_list]
column_name =  [(row.Name) for row in column_name_list]
print column_index
print column_name

df_log_reduced = df_log.select(column_index)
df_log.unpersist()
sqlContext.clearCache()

oldColumns = df_log_reduced.schema.names
newColumns = column_name
df_log_reduced = reduce(lambda df_log_reduced, idx: df_log_
reduced.withColumnRenamed(oldColumns[idx], newColumns[idx]),
xrange(len(oldColumns)), df_log_reduced)
df_log_reduced.show()

df_log_reduced = df_log_reduced.filter(df_log_reduced.
timestamp.isNotNull())
df_log_reduced.show()
print df_log_reduced.count()

df_log_reduced = df_log_reduced.withColumn("timestamp",trim(col
("timestamp")))
df_log_reduced = df_log_reduced.filter(df_log_reduced.
timestamp.isNotNull())
df_log_reduced = df_log_reduced.withColumn("timestamp", df_log_
reduced["timestamp"].cast(LongType()))
df_log_reduced.show()
print df_log_reduced.count()
```

```
db_config = sqlContext.read.format('csv').options(delimiter='
').load("/tmp/dbconfig.txt")
db_config.show()
hostname = db_config.where(db_config._c0 == "hostname").
select("_c1").rdd.flatMap(list).first()
username = db_config.where(db_config._c0 == "username").
select("_c1").rdd.flatMap(list).first()
passwd = db_config.where(db_config._c0 == "passwd").select
("_c1").rdd.flatMap(list).first()
dbname = db_config.where(db_config._c0 == "dbname").select
("_c1").rdd.flatMap(list).first()

time_threshold = sqlContext.read.format('jdbc').
options(url="jdbc:mysql://"+ hostname + '/' +
dbname,driver='com.mysql.jdbc.Driver',dbtable='threshold',
user=username,password=passwd).load()
time_threshold = time_threshold.withColumn("threshold",
time_threshold["threshold"].cast(LongType()))
threshold = time_threshold.where(time_threshold.
job == "BuildModelContent").select("threshold").rdd.
flatMap(list).first()
print threshold
print lim
df_log_reduced = df_log_reduced[df_log_reduced.timestamp >
threshold].limit(lim)
df_log_reduced.show()
print df_log_reduced.count()

status = "Regular job is continued"
if df_log_reduced.count() < lim:
    status = "Regular job is finshed"
```

```
if df_log_reduced.count() == 0:
    f = open('/home/hadoop/status.txt','w')
    f.write("Regular job is finshed")
    f.close()
    exit()

threshold = df_log_reduced.agg({"timestamp": "max"}).collect()
[0]["max(timestamp)"]
print threshold
df_log_reduced = df_log_reduced.drop("timestamp")
time_threshold1 = time_threshold.toDF("job","threshold")
time_threshold.unpersist()
time_threshold1.show()
time_threshold1 = time_threshold1.withColumn("threshold",
F.when(time_threshold1.job == "BuildModelContent",threshold).
otherwise(F.lit(0)))
time_threshold1.show()
print time_threshold1.count()
time_threshold1.write.format('jdbc').
options(url="jdbc:mysql://"+ hostname + '/' +
dbname,driver='com.mysql.jdbc.Driver',dbtable='threshold1',
user=username,password=passwd).mode('overwrite').save()

split_col = pyspark.sql.functions.split(df_log_
reduced['Identifier'], ':')
df_log_reduced = df_log_reduced.withColumn('Identifier',
split_col.getItem(3))
df_log_reduced.show()

df_log_reduced = df_log_reduced.withColumn("Node",trim(col
("Node")))
click_data = df_log_reduced.select("Node","Identifier","
NextNode")
```

215

```
click_data = click_data[ click_data.Node == ""]
click_data = click_data.drop("Node")
click_data = click_data.toDF("Identifier","Node")
#click_data.show(click_data.count(),False)
click_data.write.format('jdbc').options(url="jdbc:mysql://"+
hostname + '/' + dbname,driver='com.mysql.jdbc.
Driver',dbtable='click_info',user=username,password=passwd).
mode('append').save()

node_popular = df_log_reduced.select("NextNode","watch_time")
node_popular = node_popular.toDF("Node","watch_time")
node_popular = node_popular.groupby("Node").agg(F.
sum("watch_time"))
node_popular.show()
node_popular.write.format('jdbc').options(url="jdbc:mysql://"+
hostname + '/' + dbname,driver='com.mysql.jdbc.
Driver',dbtable='node_popular',user=username,password=passwd).
mode('overwrite').save()
node_view = df_log_reduced.select("NextNode","Identifier")
node_view = node_view.toDF("Node","Identifier")
node_view = node_view.withColumn("Identifier", F.lit(1))
node_view = node_view.groupby("Node").agg(F.sum("Identifier"))
node_view.write.format('jdbc').options(url="jdbc:mysql://"+
hostname + '/' + dbname,driver='com.mysql.jdbc.
Driver',dbtable='node_view',user=username,password=passwd).
mode('overwrite').save()

df_log_reduced = df_log_reduced.withColumn("Node",trim(col
("Node")))
df_log_reduced = df_log_reduced.withColumn("NextNode",trim(col(
"NextNode")))
df_log_reduced1 = df_log_reduced.select('Identifier','NextNod
e','Node')
```

```
df_log_reduced1 = df_log_reduced1.toDF('Identifier1','Node1',
'prevNode')
df_log_reduced = df_log_reduced.join(df_log_reduced1,((df_
log_reduced['Node'] == df_log_reduced1['Node1']) & (df_log_
reduced['Identifier'] == df_log_reduced1['Identifier1']) &(df_
log_reduced['watch_time'] != '')),'inner')
df_log_reduced1.unpersist()
sqlContext.clearCache()
df_log_reduced = df_log_reduced.drop('Node1',
'NextNode','Identifier1')
sqlContext.clearCache()
df_log_reduced.show()

df_log = df_log_reduced.withColumn("score", (F.col("watch_
time") / F.col("Duration")))
df_log_reduced.unpersist()
sqlContext.clearCache()

threshold_config = sqlContext.read.json('/tmp/config_
threshold.json')
threshold_config.show()
for row in threshold_config.collect():
    df_log = df_log.withColumn("is_watched", F.when((df_log.
    channel_id == row.channel) & (df_log.score > row.threshold)
    ,F.lit(str(1))).otherwise(F.lit(str(0))))

df_log = df_log.drop('watch_time')
df_log = df_log.drop("Duration")
df_log = df_log.drop("score")

df_log = df_log.withColumn("ip",geo_udf(df_log["ip"]))
split_col = pyspark.sql.functions.split(df_log['ip'], ',')
df_log = df_log.withColumn('city', split_col.getItem(0))
```

```
df_log = df_log.withColumn('region', split_col.getItem(1))
df_log = df_log.withColumn('country', split_col.getItem(2))
df_log = df_log.drop('ip')

df_log.show()

sqlContext.clearCache()

for co in df_log.columns:
    if str(co) != 'is_watched' and str(co) != 'Node':
        feature_prob = df_log.groupby(co).count()
        feature_prob = feature_prob.withColumn('featureCount',
        col('count'))
        feature_prob.show()
        cond_prob = df_log.groupby(['is_watched',
        str(co),'Node']).count()
        cond_prob = cond_prob.withColumn('jointCount',
        col('count'))
        cond_prob_joined = cond_prob.join(feature_prob,str(co))
        cond_prob.unpersist()
        feature_prob.unpersist()
        sqlContext.clearCache()
        cond_prob = cond_prob_joined[cond_prob_joined.is_
        watched != str(0)]
        cond_prob_joined.unpersist()
        sqlContext.clearCache()
        cond_prob = cond_prob.drop('is_watched')
        cond_prob = cond_prob.withColumn('featurevalue',  ˙
        col(str(co)))
        cond_prob = cond_prob.drop(str(co))
        cond_prob = cond_prob.drop('count')
        cond_prob = cond_prob.withColumn('featurename',
        lit(str(co)))
```

```
        cond_prob.show()
        cond_prob.write.format('jdbc').
        options(url="jdbc:mysql://"+ hostname + '/' +
        dbname,driver='com.mysql.jdbc.Driver',dbtable='score_re
        c',user=username,password=passwd).mode('append').save()
        cond_prob.unpersist()
        sqlContext.clearCache()
df_log.unpersist()
sqlContext.clearCache()
sc.stop()
f = open('/home/hadoop/status.txt','w')
f.write(status)
f.close()
```

The config files in HDFS are given here:

```
Config_channel.jsoin
{"channels" :  "moviesbyfawesome"}
{"channels" : "gousa"}
Config_drop.json
{"drop" : "watch_time"}
{"drop" : "ip"}
{"drop" : "timestamp"}
Config_parameters.json
{"Name": "Identifier", "Index" : "_1"}
{"Name": "NextNode", "Index" : "_4"}
{"Name": "timestamp", "Index" : "_6"}
{"Name": "Node", "Index" : "_20"}
{"Name": "ContentType", "Index" : "_21"}
{"Name": "Author", "Index" : "_24"}
{"Name": "Platform-Channelname", "Index" : "_7"}
{"Name": "ip", "Index" : "_9"}
```

```
{"Name": "watch_time", "Index" : "_26"}
{"Name": "Duration", "Index" : "_25"}
{"Name": "channel_id", "Index" : "_8"}
```

Config_threshold.json
```
{"channel" : "all", "threshold" : 0.5 }
{"channel" : "236", "threshold" : 0.2}
{"channel" : "922", "threshold" : 0.6}
```

Dbconfig.txt
```
hostname 10.10.10.10
username hadoop
passwd sayan123
dbname model
```

The following commands will run the code:

```
spark-submit --packages mysql:mysql-connector-java:5.1.39
build_model_first.py --driver-memory 5g --executor-memory
10g --spark.driver.maxResultSize 5g  --spark.executor.
extraJavaOptions -XX:+UseG1GC --spark.python.worker.
memory 5g --spark.shuffle.memoryFraction .5 --packages com.
databricks:spark-csv_2.10:1.1.0
```

We will discuss something crucial about the Spark memory model now. The operation is performed in memory by Spark. Using a command-line parameter, we can change the default size of execution memory. Spark can handle files that are greater than its RAM, but it must use disc memory to do it. For example, the input log file was greater than the Spark memory size, but the program successfully filters important lines from there. However, there is one exception. If you do decide to join, Spark requires the memory of both data frames. We use a join here to make our code as memory-optimized as possible by manually calling the Spark garbage collector via the unpersist function. However, it slows down the code. If you eliminate that code, the program will run faster but use more memory.

Analytics in the Cloud

Like many other fields, analytics is being impacted by the cloud. It is affected in two ways. Big cloud providers are continuously releasing machine learning APIs. So, a developer can easily write a machine learning application without worrying about the underlining algorithm. For example, Google provides APIs for computer vision, natural language, speech processing, and many more. A user can easily write code that can give the sentiment of an image of a face or voice in two or three lines of code.

The second aspect of the cloud is in the data engineering part. In Chapter 1 we gave an example of how to expose a model as a high-performance REST API using Falcon. Now if a million users are going to use it and if the load varies by much, then auto-scale is a required feature of this application. If you deploy the application in the Google App Engine or AWS Lambda, you can achieve the auto-scale feature in 15 minutes. Once the application is auto-scaled, you need to think about the database. DynamoDB from Amazon and Cloud Datastore by Google are auto-scaled databases in the cloud. If you use one of them, your application is now high performance and auto-scaled, but people around globe will access it, so the geographical distance will create extra latency or a negative impact on performance. You also have to make sure that your application is always available. Further, you need to deploy your application in three regions: Europe, Asia, and the United States (you can choose more regions if your budget permits). If you use an elastic load balancer with a geobalancing routing rule, which routes the traffic from a region to the app engine of that region, then it will be available across the globe. In geobalancing, you can mention a secondary app engine for each rule, which makes your application highly available. If a primary app engine is down, the secondary app engine will take care of things.

Figure 7-6 describes this system.

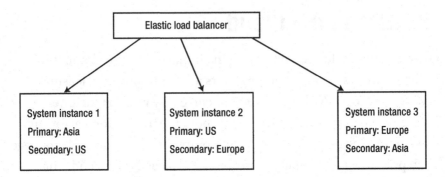

Figure 7-6. *The system*

In Chapter 1, I showed some example code of publishing a deep learning model as a REST API. The following code is the implementation of the same logic in a cloud environment where the other storage is replaced by a Google data store:

```
# Importing libraries

import falcon
from falcon_cors import CORS
import json
import pygeoip
import json
import datetime as dt
import ipaddress
import math
from concurrent.futures import *
import numpy as np
from google.cloud import datastore

# logistic function

def logit(x):
        return (np.exp(x) / (1 + np.exp(x)))

def is_visible(client_size, ad_position):
```

```
        y=height=0
        try:
                height  = int(client_size.split(',')[1])
                y = int(ad_position.split(',')[1])
        except:
                pass
        if y < height:
                return ("1")
        else:
                return ()0")
# Predictor class having all the required functions

class Predictor(object):

# constructor

        def __init__(self,domain,is_big):
                self.client = datastore.Client('sulvo-east')
                self.ctr = 'ctr_' + domain
                self.ip = "ip_" + domain
                self.scores = "score_num_" + domain
                self.probabilities = "probability_num_" + domain
                if is_big:
                        self.is_big = "is_big_num_" + domain
                        self.scores_big = "score_big_num_" + domain
                        self.probabilities_big = "probability_big_
                        num_" + domain
                self.gi = pygeoip.GeoIP('GeoIP.dat')
                self.big = is_big
                self.domain = domain
        def get_hour(self,timestamp):
                return dt.datetime.utcfromtimestamp(timestamp /
                1e3).hour
```

```
# to fetch score

    def fetch_score(self, featurename, featurevalue, kind):
        pred = 0
        try:
            key = self.client.key(kind,featurename +
            "_" + featurevalue)
            res= self.client.get(key)
            if res is not None:
                pred = res['score']
        except:
            pass
        return pred

# function to calculate score

    def get_score(self, featurename, featurevalue):
        with ThreadPoolExecutor(max_workers=5) as pool:
            future_score = pool.submit(self.
            fetch_score,featurename,
            featurevalue,self.scores)
            future_prob = pool.submit(self.fetch_score,
            featurename, featurevalue,self.probabilities)
            if self.big:
                future_howbig = pool.submit(self.
                fetch_score,featurename,
                featurevalue,self.is_big)
                future_predbig = pool.submit(self.
                fetch_score,featurename,
                featurevalue,self.scores_big)
                future_probbig = pool.submit(self.
                fetch_score,featurename,
                featurevalue,self.probabilities_big)
```

```
                pred = future_score.result()
                prob = future_prob.result()
                if not self.big:
                        return pred, prob
                howbig = future_howbig.result()
                pred_big = future_predbig.result()
                prob_big = future_probbig.result()
              return (howbig, pred, prob, pred_big, prob_big)
    def get_value(self, f, value):
        if f == 'visible':
                fields = value.split("_")
                value = is_visible(fields[0], fields[1])
                if f == 'ip':
                ip = str(ipaddress.IPv4Address(ipaddress.
                ip_address(value)))
                    geo = self.gi.country_name_by_addr(ip)
                if self.big:
                        howbig1,pred1, prob1, pred_big1,
                        prob_big1 = self.get_score('geo', geo)
                else:
                        pred1, prob1 = self.get_score('geo', geo)
                freq = '1'
                key = self.client.key(self.ip,ip)
                res = self.client.get(key)
                if res is not None:
                        freq = res['ip']
                if self.big:
                        howbig2, pred2, prob2, pred_
                        big2, prob_big2 = self.get_
                        score('frequency', freq)
                else:
```

```
            pred2, prob2 =  self.get_
            score('frequency', freq)
    if self.big:
            return ((howbig1 + howbig2), (pred1
            + pred2), (prob1 + prob2), (pred_
            big1 + pred_big2), (prob_big1 +
            prob_big2))
    else:
            return ((pred1 + pred2), (prob1
            + prob2))
if f == 'root':
    try:
            res = client.get('root', value)
            if res is not None:
                    ctr = res['ctr']
                    avt = res['avt']
                    avv = res['avv']
                    if self.big:
                        (howbig1,pred1,prob1,pred_
                        big1,prob_big1) = self.
                        get_score('ctr', str(ctr))
                        (howbig2,pred2,prob2,pred_
                        big2,prob_big2) = self.
                        get_score('avt', str(avt))
                        (howbig3,pred3,prob3,pred_
                        big3,prob_big3) = self.
                        get_score('avv', str(avv))
                        (howbig4,pred4,prob4,pred_
                        big4,prob_big4) = self.
                        get_score(f, value)
                    else:
```

```python
            (pred1,prob1) = self.get_
            score('ctr', str(ctr))
            (pred2,prob2) = self.get_
            score('avt', str(avt))
            (pred3,prob3) = self.get_
            score('avv', str(avv))
            (pred4,prob4) = self.get_
            score(f, value)
        if self.big:
            return ((howbig1 + howbig2 +
            howbig3 + howbig4), (pred1
            + pred2 + pred3 + pred4),
            (prob1 + prob2 + prob3 +
            prob4),(pred_big1 + pred_
            big2 + pred_big3 + pred_
            big4),(prob_big1 + prob_big2
            + prob_big3 + prob_big4))
        else:
            return ((pred1 + pred2 +
            pred3 + pred4), (prob1 +
            prob2 + prob3 + prob4))
    except:
        return (0,0)
    if f == 'client_time':
        value = str(self.get_hour(int(value)))
    return self.get_score(f, value)
def get_multiplier(self):
    key = self.client.key('multiplier_all_num',
    self.domain)
    res = self.client.get(key)
    high = res['high']
    low = res['low']
```

```
                if self.big:
                            key = self.client.key('multiplier_
                            all_num', self.domain + "_big")
                    res = self.client.get(key)
                    high_big = res['high']
                    low_big = res['low']
                    return(high, low, high_big, low_big)
            return( high, low)

# function to post back to ad server

    def on_post(self, req, resp):
        if True:
                input_json = json.loads(req.stream.
                read(),encoding='utf-8')
                input_json['visible'] = input_json['client_
                size'] + "_" + input_json['ad_position']
                del input_json['client_size']
                del input_json['ad_position']
                howbig = 0
                pred = 0
                prob = 0
                pred_big = 0
                prob_big = 0
                worker = ThreadPoolExecutor(max_workers=1)
                thread = worker.submit(self.get_multiplier)
            with ThreadPoolExecutor(max_workers=8) as pool:
                    future_array = { pool.submit(self.
                    get_value,f,input_json[f]) : f for f
                    in input_json}
                    for future in as_
                    completed(future_array):
                            if self.big:
```

```
                    howbig1, pred1,
                    prob1,pred_big1,prob_
                    big1 = future.result()
                    pred = pred + pred1
                    pred_big = pred_big +
                    pred_big1
                    prob = prob + prob1
                    prob_big = prob_big +
                    prob_big1
                    howbig = howbig + howbig
            else:
                    pred1, prob1 = future.
                    result()
                    pred = pred + pred1
                    prob = prob + prob1
    if self.big:
          if howbig > .65:
                  pred, prob = pred_big, prob_big
    resp.status = falcon.HTTP_200
      res = math.exp(pred)-1
    if res < 0.1:
          res = 0.1
    if prob < 0.1 :
          prob = 0.1
    if prob > 0.9:
          prob = 0.9
    if self.big:
          high, low, high_big, low_big =
          thread.result()
          if howbig > 0.6:
                  high = high_big
                  low = low_big
```

```
                else:
                        high, low = thread.result()
                multiplier = low + (high -low)*prob
                res = multiplier*res
                resp.body = str(res)
            #except Exception,e:
            #       print(str(e))
            #       resp.status = falcon.HTTP_200
            #       resp.body = str("0.1")
cors = CORS(allow_all_origins=True,allow_all_
methods=True,allow_all_headers=True)
wsgi_app = api = falcon.API(middleware=[cors.middleware])
f = open('publishers2.list_test')
for line in f:
        if "#" not in line:
                fields = line.strip().split('\t')
                domain = fields[0].strip()
                big = (fields[1].strip() == '1')
                p = Predictor(domain, big)
                url = '/predict/' + domain
                api.add_route(url, p)
f.close()
```

You can deploy this application in the Google App Engine with the following:

```
gcloud app deploy --prject <prject id>  --version <version no>
```

The function's flow is as follows:

1. The ad server requests an impression prediction.

2. When the main thread receives the request predictor, it creates a new thread for each feature.

230

3. By sending a concurrent request to the data store, each thread calculates is big, floor, floor big, probability, and probability big for each feature. It returns 0 if the feature or value is not found in the data store.

4. If is big indicates a high-value impression, large scores are used; otherwise, general scores are used.

5. It predicts score = $\sum$ score for each feature.

6. It calculates the final predicted floor using the multiplier range obtained from the data store.

7. The predicted floor value is returned to the ad server in the response.

Internet of Things

The IoT is simply the network of interconnected devices embedded with sensors, software, network connectivity, and necessary electronics that enable them to collect and exchange data, making them responsive. The field is emerging with the rise of technology just like big data, realtime analytics frameworks, mobile communication, and intelligent programmable devices. In the IoT, you can analyze data on the server side using the techniques shown throughout the book; you can also put logic on the device side using the Raspberry Pi, which is an embedded system version of Python.

Essential Architectural Patterns for Data Scientists

Data is not an isolated entity. It must gather data from some application or system, then store it accurately in certain storage, and then construct a model on top of that model, which must then be provided as an API to connect with other systems. This API must occasionally be present throughout the world with a certain latency. So, much engineering goes into creating a successful intelligent system, and in today's startup environment, which is a multibillion-dollar industry, an organization cannot afford to hire a large number of specialists to create a unique feature in their product. In the startup world, the data scientist must be a full-stack analytic expert. So, in the following sections, we'll provide different scenarios with Tom, a fictitious character, to illustrate several key architectural patterns that any data scientist should be aware of.

Scenario 1: Hot Potato Anti-Pattern

Tom is employed as a data scientist to work on a real-time analytics product for an online company. So, the initial step is to gather data from his organization's application. They use the cloud to auto-scale their storage, and they push data straight to the database from the application. In the test environment, everything appears to be in order. To ensure that there is no data loss, they use a TCP connection. When they go live, though, they do not make any changes to the main application, and it crashes. The company faces a massive loss within a half-hour, and Tom gets real-time feedback for his first step of the real-time analytic system: he is fired.

The concern now is why the main application crashes when nothing has changed. From a classic computer science perspective, this is called a *busy consumer problem*. Here the main application is the sender of data, and the database is the consumer. When the consumer is busy, which is a common occurrence in any database with a large number of queries running in it, it is difficult to handle the incoming data. Furthermore, because the TCP connection ensures the delivery data, the sender delivers the data again, which causes the sender to be loaded again, and this is the main application. The circumstance is analogous to passing a hot potato from one person to another, with the recipient returning the hot potato to the sender repeatedly. That's why it is called the Hot Potato anti-pattern. Figure 7-7 explains the sequence visually.

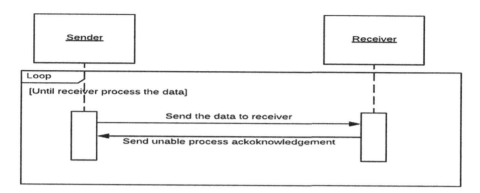

Figure 7-7. *Sequence diagram describing Hot Potato anti-pattern*

There are two elements to the situation. If the data that passes between the sender and the recipient is not required, we can use the UDP protocol, which drops the data that is unable to deliver. It's one of the factors why UDP is used in all network monitoring protocols, such as SNMP and NetFlow. It does not require the device to be loaded to monitor. However, if the data is essential, such as in the financial industry, we must create a messaging queue between the sender and the recipient. When the receiver is unable to process data, it functions as a buffer. If the queue memory is full, the data is lost, or the load is transferred to the sender. ZMQ stands for "zero message queues," and it's nothing more than a UDP socket.

In cloud platforms, there are numerous readymade solutions; we go through them in depth in Chapter 2. The following Node.js code shows an example of a collector that uses Rabit-MQ and exposes it as a REST API to the sender, using Google Big Query as the receiver.

Code: Data Collector Module

```
var https = require("https");
var express = require('express');
var router = express.Router();
var _ = require('underscore');
var moment = require('moment-timezone');
```

```
var RedisSMQ = require('rsmq');
var RSMQWorker = require('rsmq-worker');

var rsmq_h = new RedisSMQ({host: '127.0.0.1', port: 6379, ns:
'rsmq'});
rsmq_h.createQueue({qname: 'hadoopqueue'}, function (err, resp) {
  if (err) {
    console.log(err.message);
  }
  if (resp === 1) {
    console.log('✓ Message Queue created');
  }
});

var worker_h = new RSMQWorker('hadoopqueue');
worker_h.on('message', function(data, next, id) {
  const BigQuery = require('@google-cloud/bigquery');
const bigquery = BigQuery({
    projectId: 'sayantest1979'
  });
    const dataset = bigquery.dataset('adlog');
const table = dataset.table('day1');
rows = JSON.parse(data);

table.insert(rows)
    .then((insertErrors) => {
      console.log('Inserted:');
      rows.forEach((row) => console.log(row));
    });
next();
});
```

```
worker_h.on('error', function(err, msg) {
  console.log('ERROR', err, msg.id);
});

worker_h.start();
router.post('/collect', function(req, res, next) {
  res.setHeader('Access-Control-Allow-Origin', '*');

  var data = req.body,
    dataArray = [],
     for (k in data) {
    var kv = data[k].split('|');
    var dataObj = {};
    for (i=0; i < kv.length; i++) {
      var d = kv[i].split(':');
      dataObj[d[0]] = decodeURIComponent(d[1]);
    }
    dataObj['server_time'] = moment().toDate().valueOf();
    dataObj['ip'] = require('ipware')().get_ip(req).clientIp;
    var str = JSON.stringify(dataObj, require('json-decycle').
    decycle());
    dataArray.push(str);
  }
  var dataString = dataArray.join(',\n');
  dataString = "[ " + dataString + " ]";

  rsmq_h.sendMessage({qname: 'hadoopqueue', message: dataString
  + '\n'}, function (err, resp) {
      resp && console.log('Message sent. ID:', resp);
    });
    res.end('OK');
});
module.exports = router;
```

Now, let's explore other important architectural patterns, proxy and layering.

Scenario 2: Proxy and Layering Patterns

Tom starts working at a new company. There is no job uncertainty because the company is large. He does not take the risk of gathering the data in this case. The data is stored on a MySQL server. Tom had no prior knowledge of the database. He was passionate about learning MySQL. In his code, he writes lots of queries. The database is owned by another team, and their manager likes R&D. So every Monday, Tom receives a call informing him that the database has changed from MySQL to Mongo and subsequently from Mongo to SQL Server, requiring him to make modifications across the code. Tom is no longer unemployed, but he returns home from work every day at midnight.

Everyone seems to agree that the solution is to properly arrange the code. However, understanding the proxy and layering patterns is beneficial. Instead of utilizing a raw MySQL or Mongo connection in your code, employ a wrapper class as a proxy in the proxy pattern. Using the layering technique, divide your code into many layers, each of which uses a method exclusively from the layer below it. Database configuration should be done at the lowest levels, or core layer, in this instance. Above that is the database utility layer, which includes the database queries. Above that, there is a business entity layer that makes use of database queries. The following Python code will help you see things more clearly. Tom now knows that if there are any changes at the database level, he must investigate the core layer; if there are any changes in queries, he must investigate the database utility layer; and if there are any changes in business actors, he must investigate the entity layer. As a result, his life is now simple.

Database Core Layer

```
import  MySQLdb
class MysqlDbWrapper(object):
    connection = None
    cursor = None

    def __init__(self, configpath):
        f = open(configpath)
        config = {}
        for line in f:
            fields = line.strip().split(' ')
            config[fields[0]] = fields[1].strip()
        f.close()
        self.connection = MySQLdb.connect (host =
        config['hostname'], user = config['username'], passwd =
        config['passwd'], db = config['dbname'])
        self.cursor = self.connection.cursor()

    def close_mysql(self):
        self.cursor.close()
        self.connection.close()

    def get_data(self, query):
        self.cursor.execute(query)
        return self.cursor.fetchall()

------------
from pymongo import MongoClient
class MongoDbWrapper(object):
    collection = None
    client = None
    db = None

    def __init__(self,configpath):
```

```
        f = open(configpath)
        mongoconfig = {}
        for line in f:
            fields = line.strip().split('=')
            mongoconfig[fields[0]] = fields[1].strip()
        f.close()
        self.client = MongoClient('mongodb://'+mongoconfig
        ['usr']+':'+ mongoconfig['pswd']+'@'+mongoconfig['host']
        +'/'+mongoconfig['db'])
        self.db = self.client[mongoconfig['db']]
        self.collection = self.db[mongoconfig['collection']]

    def close_mongo(self):
        self.client.close()

    def get_data(self):
        return self.collection.find({"_type" : "node",
        "status":1})

Database Utility Layer
from mongo import MongoDbWrapper
import collections
import time

class MongoUtility(object):
    mongo_instance = None
    res = None

    def __init__(self, configpath):
        self.mongo_instance = MongoDbWrapper(configpath)

    def load_data(self):
        self.res = self.mongo_instance.get_data()
        #print self.res
```

```python
    def shutdown(self):
        self.mongo_instance.close_mongo()
----------
from mysql import MysqlDbWrapper

class MysqlUtility(object):
    mysql_instance = None

    def __init__(self, configpath):
        self.mysql_instance = MysqlDbWrapper(configpath)

    def load_score(self):
        scores = {}
        data = self.mysql_instance.get_data("select * from
        category_score")
        for row in data:
            feature_name= row[1].strip()
            provider = row[0].strip()
            feature_value = row[2].strip()
            score = row[4]/row[3]
            #print feature_name, feature_value, provider, score

        return scores

    def shutdown(self):
        self.mysql_instance.close_mysql()
```

The final layer will be an abstraction of a recommendation
system in JSON code like this:

```json
{
"all" : {
        "block_categories" : [916105,1011724,1011726,1011727],
        "no_of_category" : 8,
        "no_of_node" : 40,
```

```
        "no_of_match" : 2,
        "favourite_weight" : 0.5,
        "default_category" : [10037913,10037914,10053952],
        "conditional_block_categories" : {
                "rule" : [
                      {"if": {
                                "categories" : [10244361],
                                 "relation" : "or"
                      },
                      "then" : {
                                "categories" :
                                [10244361,10057506],
                                "relation" : "or",
                                "blockOrallow" : false
                      }}]
        },
        "category_weightedge":{
                "rule" :[
                              {
                              "origin" : "Identifier",
                              "weight" : 0.5
                              },
                              {
                              "origin" : "previousnode",
                              "weight" : 0.5
                              }]
        }
},
"250":{}
}
```

Thank You

We would thank you for reading this book. We hope you have enjoyed reading the book as we much as enjoyed writing it. We hope it helps you make a footprint on machine learning community. We would like to encourage you to keep practicing on different data sets and contribute to the open-source community. You can always contact us through the publisher. Thank you once again and best wishes!

Index

Printed in the United States
by Baker & Taylor Publisher Services